WETLANDS ON THE EDGE

The Future of Southern California's Wetlands

REGIONAL STRATEGY 2018

SOUTHERN CALIFORNIA
**WETLANDS
RECOVERY**
PROJECT

SUGGESTED CITATION

Southern California Wetlands Recovery Project. 2018. Wetlands on the Edge: The Future of Southern California's Wetlands: Regional Strategy 2018. Prepared by the California State Coastal Conservancy, Oakland, Ca.

VERSION

2018 (v1.0)

REPORT AVAILABILITY

To access the *Regional Strategy 2018,* and associated maps and tools, visit scwrp.databasin.org.

For inquiries and information on the Wetlands Recovery Project, visit scwrp.org.

FUNDED BY

California State Coastal Conservancy

U.S. Environmental Protection Agency, Wetland Program Development Grant Program

U.S. Fish and Wildlife Service, California Landscape Conservation Cooperative

National Oceanic and Atmospheric Administration National Centers for Coastal Ocean Science

REGIONAL STRATEGY AUTHORS

Editors

Megan Cooper	California State Coastal Conservancy
Evyan Sloane	California State Coastal Conservancy

Primary Authorship

Jeremy Lowe	San Francisco Estuary Institute
Eric Stein	Southern California Coastal Water Research Project

Contributing Authors

Erin Beller	San Francisco Estuary Institute
Josh Collins	San Francisco Estuary Institute
Megan Cooper	California State Coastal Conservancy
Jeff Crooks	Tijuana River National Estuarine Research Reserve
Heather Dennis	San Francisco Bay Conservation and Development Commission
Cheryl Doughty	University of California, Los Angeles
Greg Gauthier	California State Coastal Conservancy
Julie Gonzalez	California State Coastal Conservancy (California Sea Grant Fellow)
Kerstin Kalchmayr	California State Coastal Conservancy (California Sea Grant Fellow)
Shawn Kelly	Earth Island Institute
Jeremy Lowe	San Francisco Estuary Institute
Katie McKnight	San Francisco Estuary Institute
Amy Richey	San Francisco Estuary Institute
April Robinson	San Francisco Estuary Institute
Evyan Sloane	California State Coastal Conservancy
Eric Stein	Southern California Coastal Water Research Project
Martha Sutula	Southern California Coastal Water Research Project
Christine Whitcraft	California State University, Long Beach

Design and Production

Ruth Askevold	San Francisco Estuary Institute
Katie McKnight	San Francisco Estuary Institute
Julie Gonzalez	California State Coastal Conservancy (California Sea Grant Fellow)

WETLANDS RECOVERY PROJECT PARTICIPANTS

Director's Group

State Partners

John Laird, Secretary (Chair of the Directors Group)	California Natural Resources Agency
John Ainsworth, Executive Director	California Coastal Commission
Jon Bishop, Chief Deputy Director	State Water Resources Control Board
Charlton Bonham, Director	California Department of Fish & Wildlife
John Donnelly, Executive Director	Wildlife Conservation Board
David Gibson, Executive Officer	San Diego Reg. Water Quality Control Board
Jennifer Lucchesi, Executive Officer	State Lands Commission

John Largier	University of California, Davis
Shelley Luce	Environment Now
Eric Stein	So. California Coastal Water Research Project
Martha Sutula	So. California Coastal Water Research Project
Christine Whitcraft	California State Long Beach University

Wetland Advisory Group Members

SAN DIEGO COUNTY

Eric Bowlby	San Diego Canyonlands
Brian Collins	U.S. Fish and Wildlife Service
Doug Gibson	San Elijo Lagoon Conservancy
Mike Hastings	Los Peñasquitos Lagoon Foundation
Chris Nordby	Nordby Biological
Chris Peregrin	California State Parks
Jim Peugh	San Diego Audubon Society

ORANGE COUNTY

Dennis Baker	Orange Coast River Park
Vic Leipzig	Amigos de Bolsa Chica
Krista Sloniowski	Newport Bay Conservancy
George Sutherland	Trout Unlimited
Matt Yurko	Coastal Commission
Dick Zembal	Orange County Water District

LOS ANGELES COUNTY

David Cannon	Everest Consultants
John Mack	Catalina Island Conservancy
Rosi Dagit	RCD of the Santa Monica Mountains
Stacie Smith	NOAA Restoration Center
Melina Watts	Santa Monica Mountains Watershed Council
Chris Webb	Moffat & Nichol

VENTURA COUNTY

Derek Poultney	Ventura Land Trust
Laura Riege	The Nature Conservancy
Valerie Vartanian	Mugu Lagoon, Naval Base Ventura County

SANTA BARBARA COUNTY

Tom Dudley	Marine Science Institute, UCSB
Mauricio Gomez	South Coast Habitat Restoration
David Hubbard	Coastal Restoration Consultants
George Johnson	City of Santa Barbara
Adam Lambert	Marine Science Institute, UCSB
Ken Owen	Channel Islands Restoration
Lisa Stratton	UCSB Natural Reserve
Bob Thiel	Retired California State Coastal Conservancy Staff

The rapid urbanization of Southern California over the last 150 years has cost the region nearly all of its coastal wetlands. What remains are some of Southern California's most priceless natural spaces. These sensitive habitats are extremely vulnerable to the stresses of development, climate change and sea level rise. The Southern California Wetlands Recovery Project has worked for nearly 20 years to protect and expand Southern California's remaining wetlands; this Regional Strategy Update lays out the roadmap for the next 20, using a science-based approach to ensure the long-term survival of these critical habitats. I applaud the Wetlands Recovery Project on a comprehensive and forward-thinking strategy for immediate and courageous action.

— John Laird
CALIFORNIA SECRETARY
FOR NATURAL RESOURCES

CONTENTS

THE SOUTHERN CALIFORNIA WETLANDS RECOVERY PROJECT

PROJECT MISSION:

To expand, restore and protect wetlands in Southern California's coastal watersheds.

PROJECT VISION:

Restored and protected wetlands and rivers along the Southern California coast benefitting wildlife and people.

ORMOND BEACH • PHOTO COURTESY OF CALIFORNIA STATE COASTAL CONSERVANCY

INTRODUCTION
to the *Regional Strategy 2018*

WHO WE ARE

The Southern California Wetlands Recovery Project (WRP) is a partnership of 18 State and Federal agencies, chaired by the California Resources Agency and supported by the California State Coastal Conservancy. Through the WRP partnership, public agencies, scientists, and local communities work cooperatively to acquire and restore wetlands in coastal Southern California. The WRP uses a non-regulatory approach by coordinating with agency partners, although many of the member agencies implement their own regulatory mandates.

BOLSA CHICA ECOLOGICAL RESERVE • PHOTO BY SERGEI GUSSEV, COURTESY OF CREATIVE COMMONS

By cultivating resilient wetlands on a landscape scale, the WRP aims to enhance the economic, environmental and recreational benefits of wetlands in Southern California. The WRP's partners have completed 206 wetlands projects since 1999, leading to the acquisition of 8,246 acres of land and the restoration of 4,884 acres of wetlands.

The WRP was created in 1997 as a regional voice for the valuable coastal wetland resources of Southern California. Prior to the creation of the WRP, limited regional coordination or communication existed among public agencies, nonprofit organizations, and community members who had interest in Southern California's wetlands. Through the WRP's visionary regional approach, individual local efforts can be coordinated to accomplish regional and larger scale goals.

Representatives from each of the WRP partner agencies form the WRP **Directors Group** and **Managers Group**, which are made up of staff from each of the partner agencies (Figure 1). The **Wetland Advisory Group** provides local input from on-the-ground land managers and restoration practitioners. The **Science Advisory Panel** provides science recommendations to other WRP groups. The **County Task Forces** represent the WRP community at large and are made up of stakeholders and practitioners who are called upon on an as-needed basis to help identify on-the-ground issues, promote wetlands education, and implement projects.

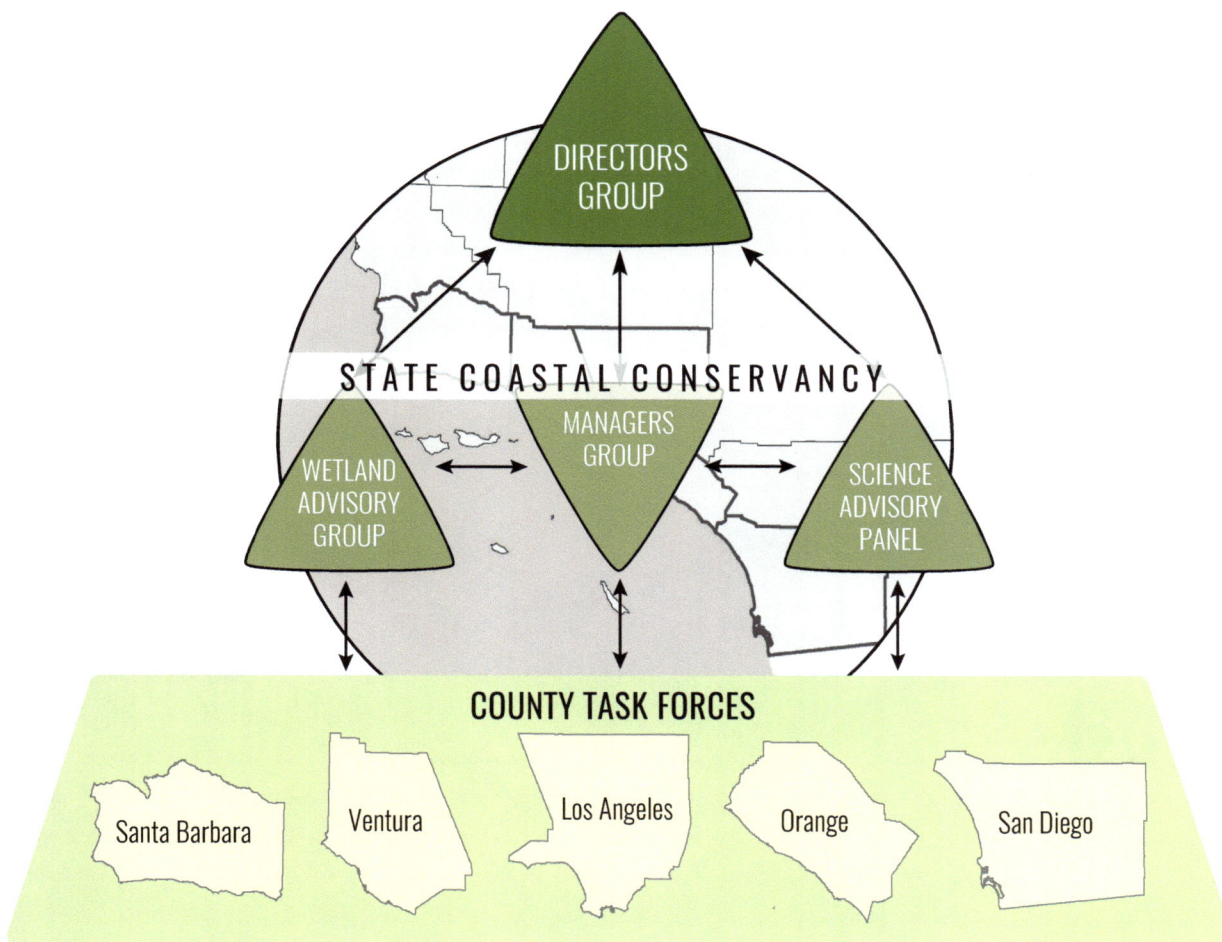

Figure 1. Organizational chart for the Wetlands Recovery Project.

WHY WETLANDS?

Throughout most of recorded world history, wetlands were regarded as wastelands and problem areas to be drained and filled. Despite this history, a shift in the understanding and appreciation for these habitats has occurred, and wetlands are now valued worldwide for the many benefits they provide (Needles et al. 2015; Costanza et al. 1997) (Figure 2). The Millennium Ecosystem Assessment (Millennium Ecosystem Assessment 2005) identifies 18 ecosystem services attributed to coastal wetlands. The WRP created a paired-down list of ecosystem services identified as the most important to the WRP (Appendix 1). Those services were deemed important based on restoration priorities for coastal wetland restoration projects in the region. The WRP's highest priority services include conserving native wildlife, carbon storage, improvement of water quality, flood and shoreline protection, aesthetic value, controlling disease and vectors, and providing opportunities for recreation and for science and education (Figure 2). Coastal wetlands provide habitat for plants and animals, including many unique and threatened or endangered species, and serve as critical fish nursery areas. The drastic loss of wetland form and function has spawned an era of much-needed wetland restoration.

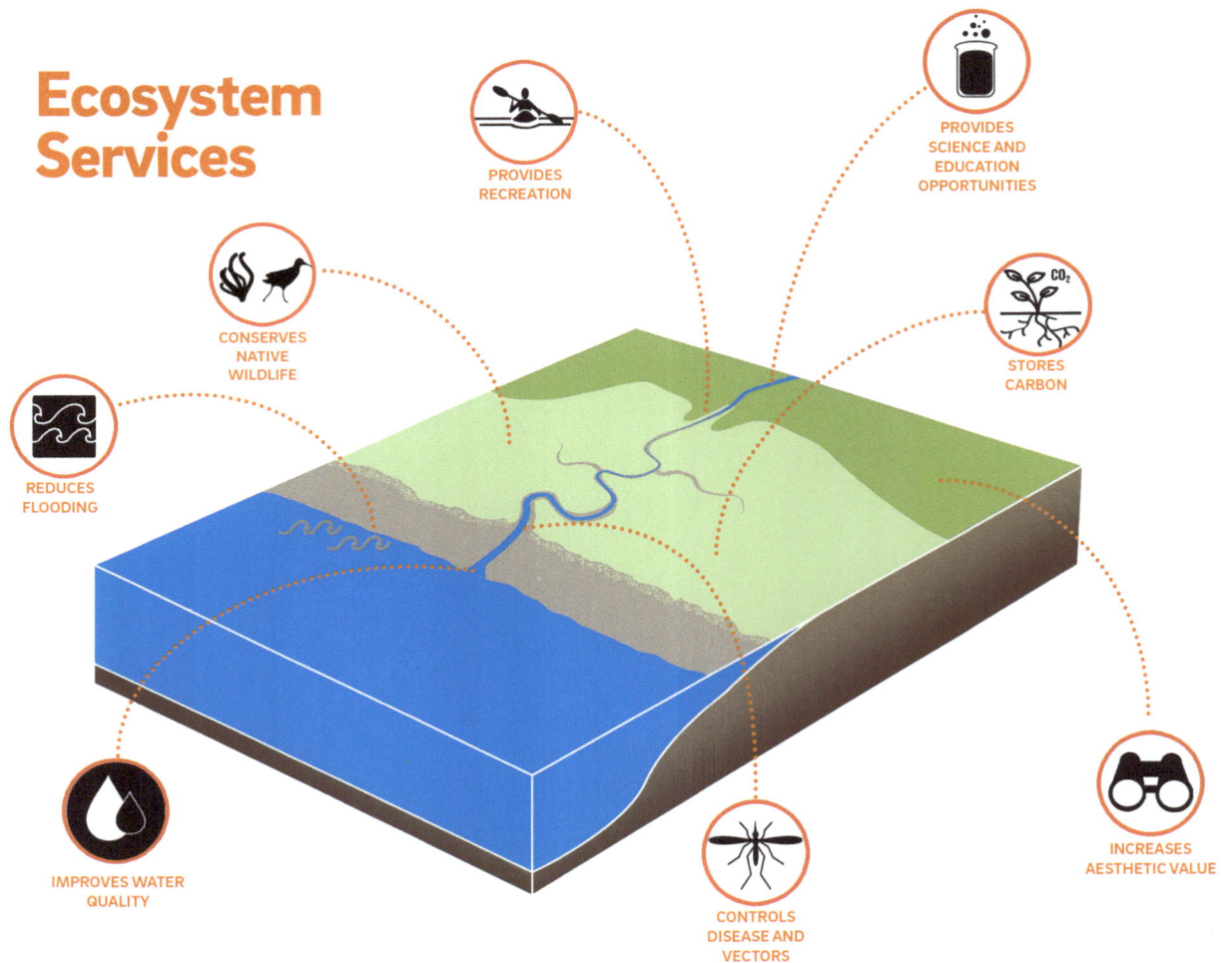

Ecosystem Services

PROVIDES RECREATION

PROVIDES SCIENCE AND EDUCATION OPPORTUNITIES

CONSERVES NATIVE WILDLIFE

STORES CARBON

REDUCES FLOODING

IMPROVES WATER QUALITY

CONTROLS DISEASE AND VECTORS

INCREASES AESTHETIC VALUE

Figure 2. Infographic of the 8 ecosystem services identified by the WRP for Southern California's coastal wetlands.

WETLAND DEFINITIONS

The WRP does not define wetlands based on a particular regulation or jurisdiction. In this document, we will use the following terms for the wetlands and habitats that we discuss:

Coastal wetlands: These are the coastal tidal wetland ecosystems that include shallow subtidal and intertidal channels, vegetated marsh, unvegetated flats, and adjacent upland transitional areas. These include coastal wetlands fringing areas of open bays and harbors, such as Mission Bay and San Diego Bay. The deeper present-day subtidal areas of open bays and harbors are not included.

Within coastal wetlands, the following habitats and zones have been defined:

Subtidal includes all areas within the wetland that are inundated year-round such as deep and shallow subtidal areas and wetland channels. Subtidal can include habitats such as rock bottoms, unconsolidated bottoms, submerged aquatic vegetation beds, or reefs (e.g., oyster reefs).

Unvegetated flats include both tidal and supratidal flats, such as mudflats, sand flats, and salt flats.

Vegetated marsh includes low (intertidal), mid- and high marsh elevation zones typically dominated by cordgrass (low zones) and pickleweed (mid-to-high zones) vegetation.

Wetland-upland transition zones include non-tidal habitats adjacent to a coastal wetland edge up to 1,600 feet wide, that encompass the ecosystem functions and services associated with the wetlands, and can include habitats such as alkali wetlands, riparian areas, coastal sage scrub, and many other upland habitats.

Non-tidal wetlands These wetlands include freshwater marshes, vernal pools, slope and seep wetlands, lakes, and non-tidal flats in the coastal watersheds.

Streams and adjacent habitats The stream, its floodplain, and additional upland buffer habitat.

FRAMEWORK OF THE *REGIONAL STRATEGY 2018*

If the individual wetlands in Southern California are considered pieces of a puzzle that we are trying to assemble, then this *Regional Strategy 2018* provides us with the picture on the back of the puzzle box. This picture will provide restoration practitioners, project proponents, and WRP agencies with a regional ecological context within which to make project-specific decisions and allow us to consider broader ecological connections and functions across wetlands. It will also allow us to better consider how to allocate limited resources across the region instead of evaluating each wetland project in isolation. The *Regional Strategy 2018* will provide the guidance for the WRP and its stakeholders to achieve the WRP's four Goals.

This document is an update of the *Regional Strategy 2001*. The *Regional Strategy 2001* has guided the collaborative efforts of the WRP agencies for over 15 years, but the need for an update was identified early on. In a May 2002 position paper, the Science Advisory Panel (SAP) states that, in reviewing the *Regional Strategy 2001*, they realized the need to "better articulate the major elements of wetland ecosystem structure and function that must be recovered in order to ultimately achieve the guiding vision and programmatic regional goals." Additionally, the *Regional Strategy 2001* did not consider the issue of sea-level rise, nor did it consider potential objectives to address it.

This *Regional Strategy 2018* is written within the following conservation framework (Figure 3). The WRP's Vision, Mission, and Goals articulate the collective approach of the WRP agencies and partners. A set of Guiding Principles provide criteria for each restoration project. The **Quantitative Objectives** (the "Objectives") are the primary building blocks of the *Regional Strategy 2018*. They provide numeric targets that will help quantify progress towards meeting the WRP's Goals and realizing the Vision. The WRP developed Objectives and Management Strategies for each of the four Goals.

The *Regional Strategy 2018* is primarily focused on developing Objectives for Goal 1 for coastal wetlands. Although Goal 1 is the primary focus of this document, Objectives for Goal 2 for non-tidal wetlands will help the WRP accomplish Goal 1 and its associated Objectives by addressing watershed conditions and restoring streams, adjacent habitat, and other non-tidal wetlands. The WRP has long recognized the need to engage with and support the local community in order for the WRP to realize its Vision, which lead to the Objectives for Goal 3. Throughout the development of the *Regional Strategy 2018*, the WRP has identified many remaining scientific questions, which have been captured in the Objectives for Goal 4.

VISION
Restored and protected wetlands and rivers along the Southern California Coast benefitting wildlife and people

MISSION
The Southern California Wetlands Recovery Project aims to expand, restore and protect wetlands in Southern California's coastal watersheds.

GUIDING PRINCIPLES (17)

GOAL 1
Preserve and restore resilient coastal tidal wetlands and associated marine and terrestrial habitats.

GOAL 2
Preserve and restore streams, adjacent habitats, and other non-tidal wetland ecosystems to support healthy watersheds.

GOAL 3
Support education and compatible access related to coastal wetlands and watersheds.

GOAL 4
Advance the science of wetland restoration and management in Southern California.

Quantitative Objectives
Management Strategies

Quantitative Objectives
Management Strategies

Quantitative Objectives
Management Strategies

Quantitative Objectives
Management Strategies

Figure 3. The conservation framework of the Southern California Wetlands Recovery Project.

KEY CONCEPTS OF THE OBJECTIVES

Throughout the development of the Objectives, the WRP developed and adhered to the following key concepts:

1) **These Objectives are based on the concept of working with nature to restore natural processes.** Physical, biological, and chemical processes shape landscapes and habitats. With their natural spatial and temporal variability, natural process create complex, heterogeneous landscapes and habitats. They help to determine which ecological functions are likely to persist and whether species will be able to adapt to environmental change in a particular place. Thus, basing restoration on an understanding of natural processes for a particular place will support a more functional and resilient wetland ecosystem.

2) **These Objectives are based on the concept that wetlands are complete systems.** Wetlands are not a particular habitat or vegetative type, but instead include many natural processes and habitat types working together from the subtidal to the upland edge, and from the stream channel to the adjacent riparian habitat. In Southern California, wetlands that we today consider individual sites are often fragments of a larger, previously connected system.

3) **These Objectives are based on the concept that working with nature to restore natural processes in whole wetland systems will support all 17 ecosystem services** identified in the Millennium Ecosystem Assessment (Millennium Ecosystem Assessment 2005).

4) **These Objectives are aspirational.** The WRP set a high target for restoration in order to provide the most resilient wetland recovery possible. The Objectives are based on the best available scientific recommendations to recover the ecological functions of coastal wetlands. The resulting quantitative Objectives and Management Strategies may seem unattainable at a particular location, but we assume that constraints on their application will be added during site-specific planning and implementation.

5) **These Objectives provide a regional perspective and may not be appropriate for every individual system.** Southern California's wetlands are more than the sum of their parts. They exist in a connected matrix of urban and natural landscapes, and the best approaches for protecting and restoring them must be considered at a regional scale. The regional Objectives can be scaled down to site-specific decisions through consideration of ecological trade-offs at the given site and its conditions and constraints.

6) **These Objectives were developed at three spatial scales: wetland, subregion, and region.** Development of Objectives at multiple spatial scales allows for consideration of physical and biological process interactions within and between individual wetlands, and for the promotion of more resilient landscapes.

The **Regional Strategy 2018** is based on:

the *PAST*

The term "historical" refers to the landscape as it looked circa 1850, as surveyed by the United States Coast Survey of the mid-to-late 19th century ("T-sheets") and site-specific historical ecology studies. Using historical information does not imply that the goal of restoration is to return a wetland to precisely the structure and composition it had in the past. Due to irreversible changes to the landscape, that may not be possible or desirable given current ecological, social, and economic considerations. Instead, historical information provides us with the best information about what wetlands were like before the changes brought about by Euro-American settlement. Restoration is about "learning how to discover the past and bring it forward into the future" (Egan and Howell 2001).

the *PRESENT*

Mapping current wetland distribution and conditions in the region provides us with the opportunities and constraints of our modern landscape. The "present" condition is forever changing due, for instance, to tidal wetland restoration projects by the coast, or water diversions/recycled water projects in the watersheds. The term "present" refers to the most recent wetland mapping available, as represented by the most recent National Wetland Inventory (NWI) (Stein et al. 2014).

and the *FUTURE*

Strategies for restoration and adaptation of wetlands impacted by climate change will provide a roadmap for the survival of these ecosystems. For the coastal wetlands, the term "future" is considered in light of rising sea levels, as defined by the recommended National Research Council (NRC) sea-level projections for the Los Angeles area (Table 5 in NRC 2012): 60.8 centimeters for the year 2050 (24-inch scenario) and 166.5 centimeters for the year 2100 (66-inch scenario). The lower, near-term projection has been used to develop restoration Objectives; the higher, longer-term projection has been used to characterize potential changes (Appendix 3). For the non-tidal wetlands, potential habitat changes due to climate change-induced alteration of rainfall patterns is much less predictable than sea-level rise, and regional projections have not yet been developed.

GUIDING PRINCIPLES

The following Guiding Principles convey the key restoration priorities and approaches of the WRP, and provide guidance for individual project development and review. Every project that supports the Goals and Objectives of the *Regional Strategy 2018* should consider these Guiding Principles from design to implementation.

1) Actions to protect and restore wetland ecosystems and adjacent habitat types support a mosaic of functional wetlands and provide habitat connectivity among wetlands, within watersheds, and along the Pacific Flyway.

2) Actions that influence the distribution of wetland archetypes consider the historic, current, and possible future extent, diversity and relative proportion of wetland types.

3) Projects have clear environmental goals that include quantifiable measures of success, and are based on scientific evaluation of feasible alternatives.

4) Projects restore and preserve ecological and physical processes to maximize ecosystem benefits based on the best available evidence for historical, present, and future conditions.

5) Projects preserve and restore the suite of locally appropriate native wetland habitats and associated species communities, including special status species.

6) Projects develop and include an adaptive management plan that outlines monitoring thresholds to trigger specific management actions.

7) Restoration results in wetland systems that are resilient to sea-level rise and other climate change stressors.

8) Projects set out to reduce key stressors to the system such as removing infrastructure barriers to hydrology or reducing watershed pollution.

9) Restoration of wetlands minimizes the scale, frequency, and cost of maintenance and long-term management.

10) Projects demonstrate incorporation and application of best-available science and lessons learned from past and present projects.

11) Projects demonstrate an explicit evaluation of ecological trade-offs.

12) Projects demonstrate an evaluation of financial costs and benefits.

13) Monitoring of projects include consistent protocols that assess project success and regional progress, allow for analysis and a statewide comparison of monitoring results.

14) Projects support wetland-associated ecosystem services.

15) The Wetland Recovery Project and associated projects share information, engage stakeholders and community members, and provide opportunity for participation.

16) Projects include public access, recreation and education opportunities, and public communication where appropriate to complement preservation of wetlands.

17) The Wetland Recovery Project actively engages, as appropriate, in the development of funding strategies, planning, and policies that promote the Wetland Recovery Project's Vision.

IMPLEMENTING THE *REGIONAL STRATEGY*

The *Regional Strategy 2018* will be implemented through the work of the WRP partners and updated as we make progress toward our Objectives and as new science becomes available. The Objectives will help guide all levels of stakeholders in the wetlands community from resource agencies and funders, to restoration practitioners designing projects, reviewing project proposals, and making funding decisions. Restoration practitioners and land managers will utilize the project guidance and regional data presented in this document to develop projects that accomplish the WRP Goals and Objectives.

The WRP **Work Plan** is a list of projects that are consistent with the Goals, Objectives, Management Strategies, and Guiding Principles identified in the *Regional Strategy 2018*. Projects are added to the Work Plan through an application and vetting process that involves review by the Wetland Managers Group and adoption by the Directors Group. The *Regional Strategy 2018* provides the WRP agencies with a framework to discuss, assess and provide feedback on projects. The Work Plan allows the funding agencies to agree on project design and approach, and coordinate funding for the most efficient and effective expenditure of resources.

The Work Plan identifies projects that meet all four of the WRP's Goals, and funding for projects will come through the unique funding sources of each WRP partner agency. Work Plan projects range from flagship tidal wetlands restoration projects to scientific studies focused on improving our knowledge of wetlands restoration and management. Stream restoration and fish passage projects are essential to the Work Plan, as is the Community Wetlands Restoration Grant Program, which focuses on education and outreach while also restoring habitat.

> ## MARSH ADAPTATION PLANNING TOOL
> The WRP has created the Marsh Adaptation Planning Tool (MAPT at scwrp.databasin.org) to assist stakeholders in developing Work Plan proposals and accessing the data from the Regional Strategy.

DESCRIPTION OF THE REGION

The physical features, climate, and hydrology of coastal Southern California have produced an unusual set of conditions and a diversity of plants and animals that distinguish the region from any other in North America. Unlike the broad, gradually sloping coastal plains of the Atlantic and Gulf Coasts, Southern California has steep, coastal mountains that descend sharply to the ocean. Summers are hot and dry in this semi-arid, Mediterranean climate, while winters are cool with rainfall varying in amount and intensity, from droughts to steady rains to torrential downpours. For instance, the San Gabriel and San Bernardino Mountains can experience more rain in a twelve-hour period than anywhere else in the continental United States.

Five Subregions

For the *Regional Strategy 2018*, the Southern California Bight has been divided into five subregions - Santa Barbara, Ventura, Santa Monica, San Pedro and San Diego (Figure 4). These subregions generally reflect the change between steep, terraced, and flatter prograding stretches of coast (Jacobs et al. 2011). They also reflect changes in orientation from southerly facing, with lower exposure to waves, to westerly facing, with higher exposure to waves. The inland boundaries of the subregions has been defined by watershed boundaries (Appendix 2).

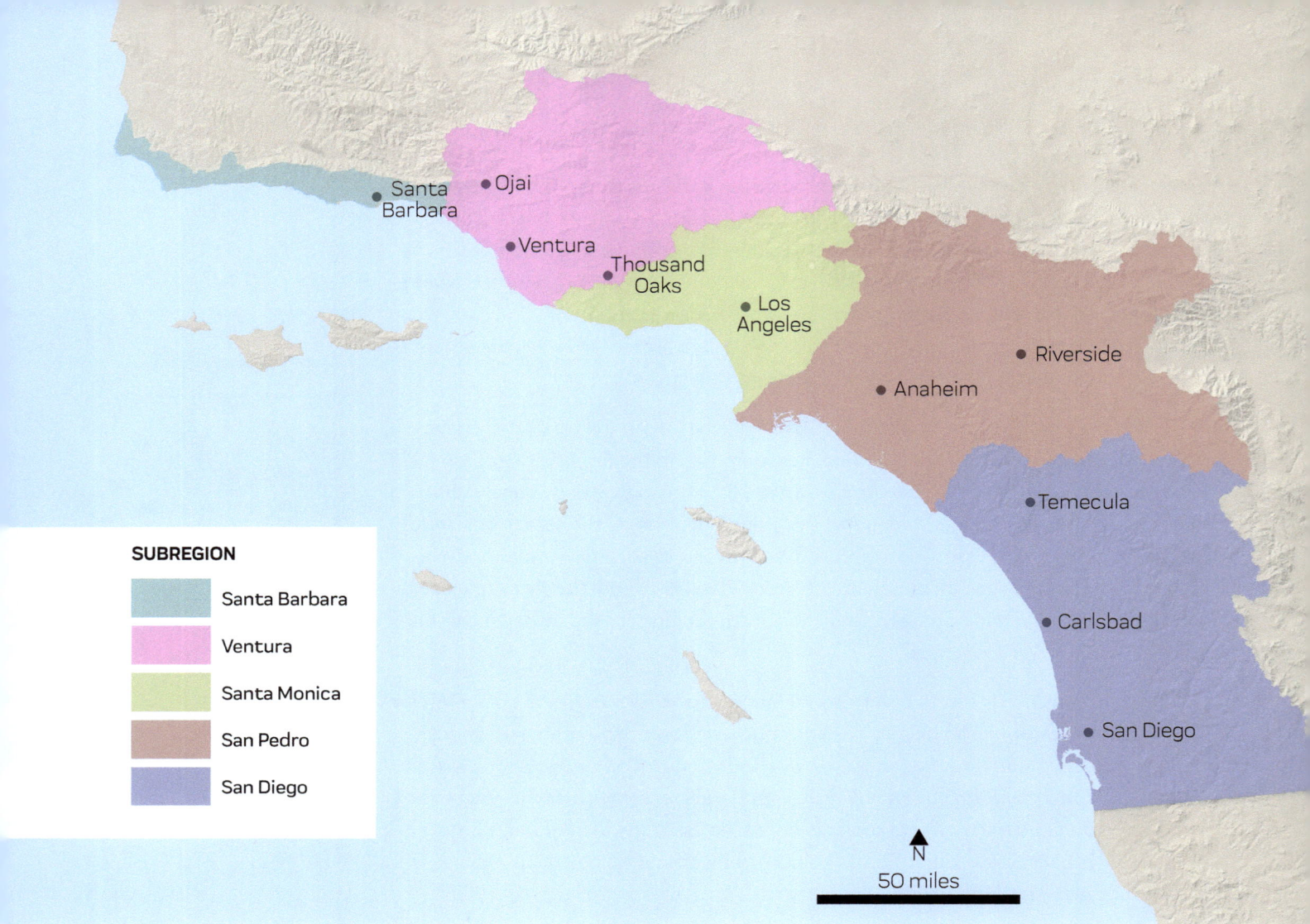

Figure 4. Extent of the five RSU subregions of Santa Barbara, Ventura, Santa Monica, San Pedro and San Diego.

SUBREGION
- Santa Barbara
- Ventura
- Santa Monica
- San Pedro
- San Diego

EAST BLUFF, UPPER NEWPORT BAY • PHOTO BY SERGEI GUSSEV, COURTESY OF CREATIVE COMMONS

The **wetland scale** represents an entire coastal wetland, where all the tidal and intertidal areas drain through a common inlet to the ocean. Wetlands can be comprised of one or multiple wetland types and can contain various habitat types. The **subregion scale** (Figure 4) reflects the major geomorphic processes which drive the coastal landscape: topography, relative exposure to waves, and size of the watershed. The **region scale** refers to the whole Southern California Bight, from Point Conception to the Mexican border. ●

COLORADO LAGOON RESTORATION, CITY OF LONG BEACH • PHOTO BY ERIC LOPEZ, CALIFORNIA COASTAL COMMISSION

GOAL 1: Restore Coastal Wetlands

The first Goal of the *Regional Strategy 2018* is to *preserve and restore resilient coastal tidal wetlands and associated marine and terrestrial habitats*. Coastal tidal wetlands, or "coastal wetlands," include all estuaries, lagoons and other wetlands that have oceanic influences.

CLASSIFICATION OF COASTAL WETLANDS

Coastal Wetland Archetypes

There are 105 wetlands addressed in the *Regional Strategy 2018*, encompassing a variety of wetland types—lagoons and river mouths, large and small (Figure 5). Developing Objectives for all of these wetlands requires some level of generalization, or else each recommendation would be site-specific. As stated in Key Concept 2 (page 8), the Objectives are based on complete

wetlands, including multiple habitats and processes. As such, we need a classification that captures the whole wetland system. Many of the 105 individual wetlands share some common characteristics, which can be used to create recommendations that are generalizable across the region. The *Regional Strategy 2018* employs a wetland classification system that creates groups of wetlands that are based on processes and functions rather than habitat types. This classification is called an "Archetype."

An archetype is a group of wetlands that are similar in terms of form, function, and processes. The physical conditions used to develop the archetype classifications include catchment properties (levels of water and sediment inputs), wetland area, proportion of subtidal and intertidal area, inlet dimension and condition, and tidal volume. The archetypes provide a general model that can be used to explain how a group of wetlands functions, and how those wetlands may respond to external pressures or drivers. The WRP defined seven archetypes that represent the range of tidal wetlands across the Bight and are presented below (Figures 6-10). Detailed archetype descriptions and classifications of every wetland can be found in Appendix 2.

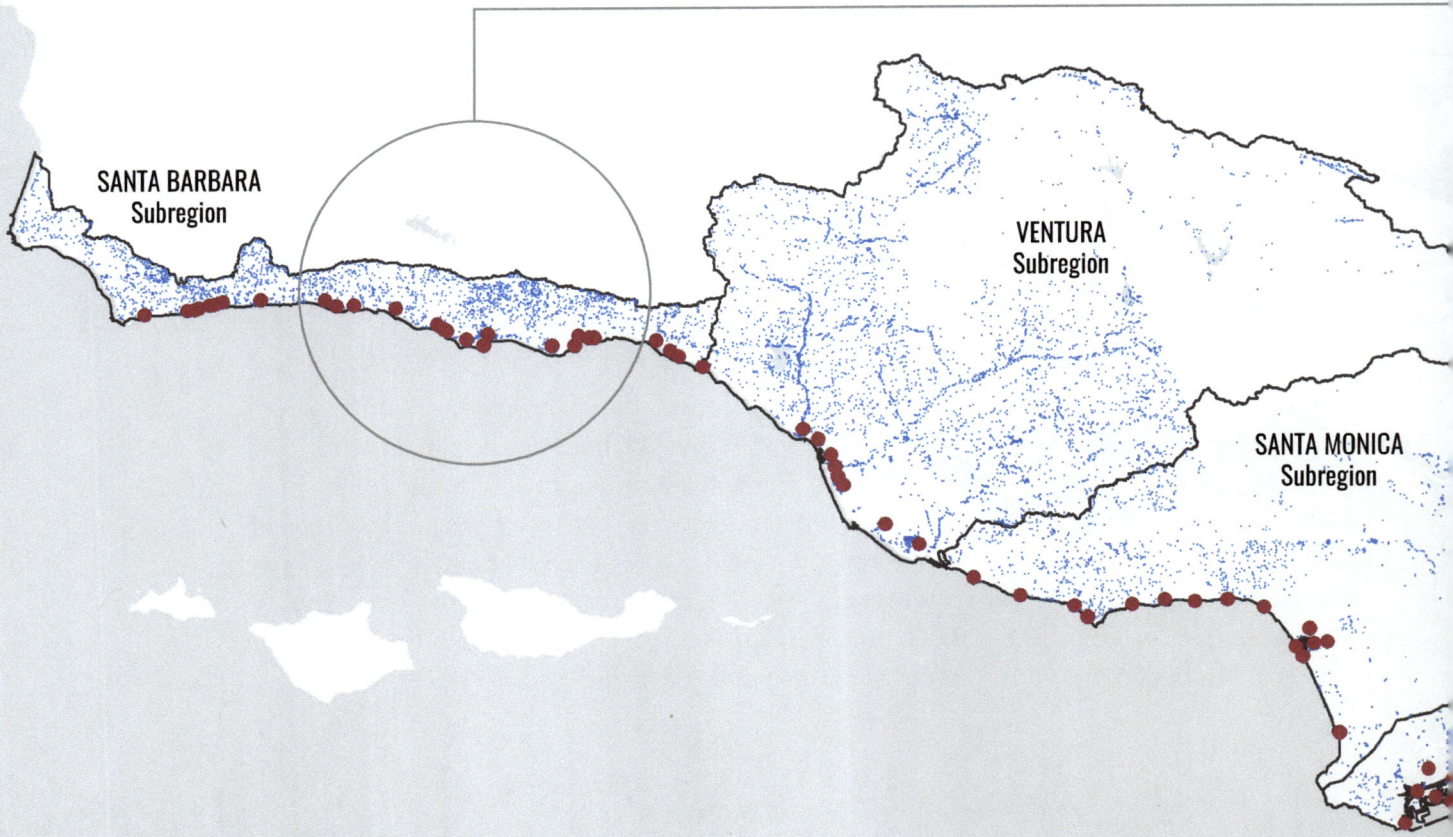

WETLANDS ACROSS THE WRP SUBREGIONS

● Coastal Wetlands

Non-Tidal Wetlands

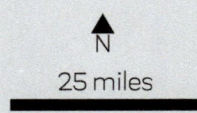

N

25 miles

Non-Tidal Wetlands

Coastal Wetlands

SAN PEDRO
Subregion

SAN DIEGO
Subregion

UNITED STATES
MEXICO

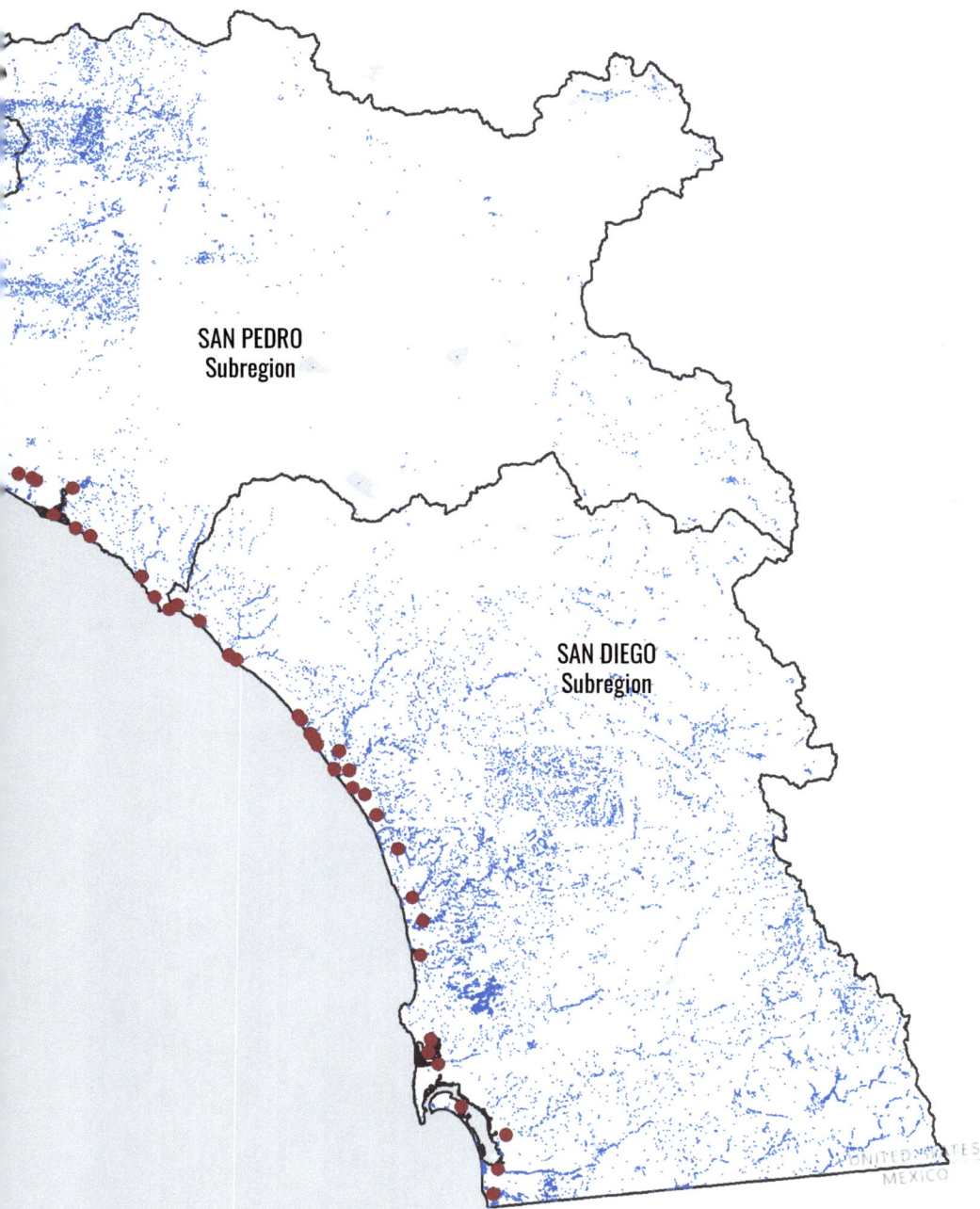

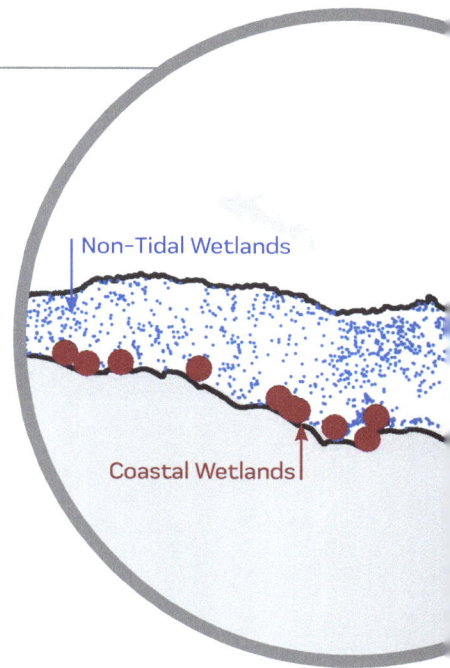

Figure 5. Wetlands across the WRP subregions. Each coastal wetland is indicated by a red dot. Non-tidal wetlands are shown as individual blue dots, indicating the center of that wetland feature. An enlargement of the map is shown above.

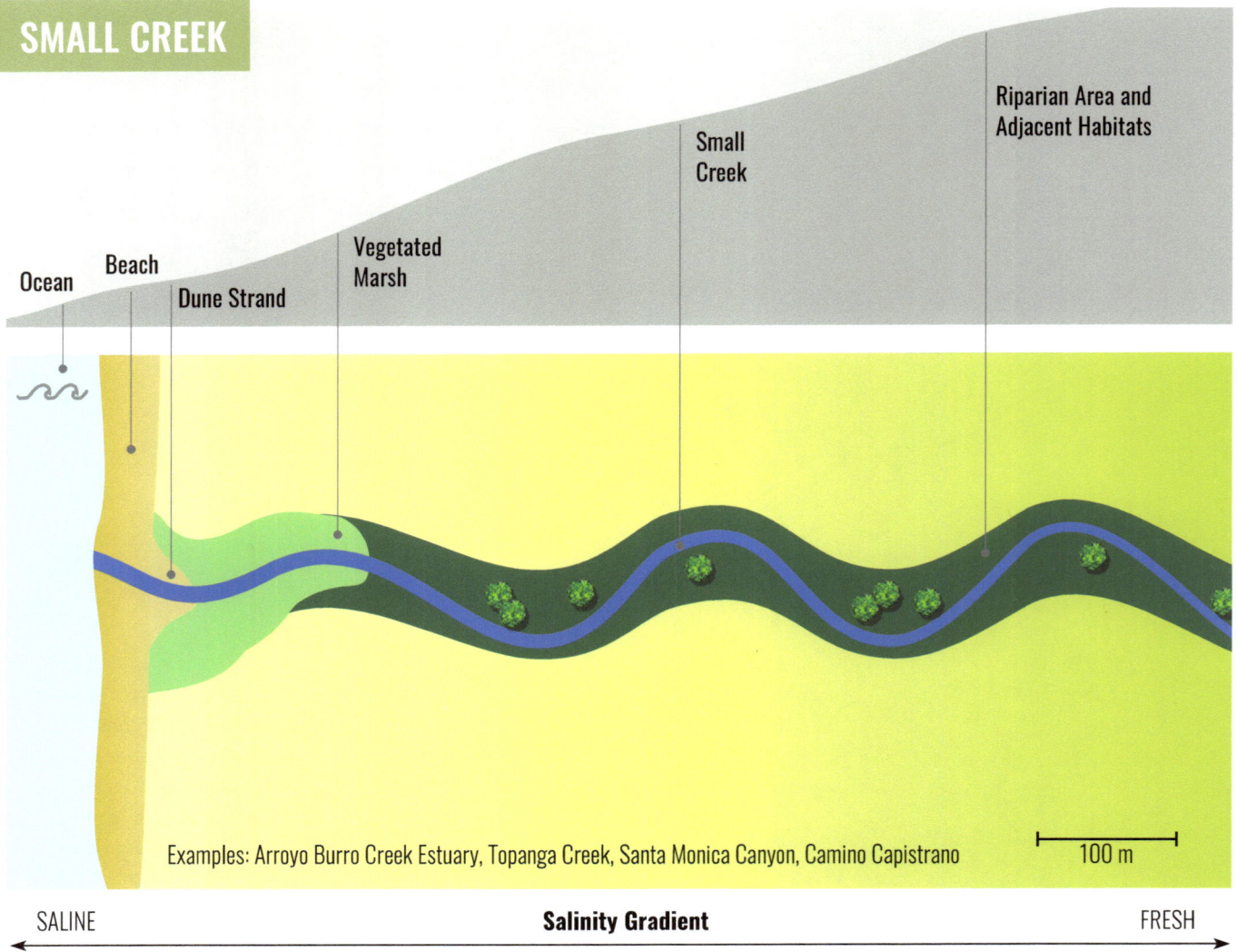

Ocean Beach Dune Strand Vegetated Marsh Small Creek Riparian Area and Adjacent Habitats

Examples: Arroyo Burro Creek Estuary, Topanga Creek, Santa Monica Canyon, Camino Capistrano

100 m

SALINE ← **Salinity Gradient** → FRESH

Figure 6. Diagram of the small creek archetype.

Definition of small creek: A small inlet with minimal subtidal habitat area, a small area of vegetated marsh at the inlet, and a generally steeper channel slope. Steep watersheds and narrow valleys control the size of the creeks and the area available for wetlands. These creeks are intermittently open to the ocean throughout the year, creating an estuarine salinity gradient during open periods.

EXAMPLE OF SMALL CREEK: ARROYO BURRO CREEK ESTUARY • COPYRIGHT (C) 2002-2018 K. AND G. ADELMAN, CALIFORNIA COASTAL RECORDS PROJECT, WWW.CALIFORNIACOASTLINE.ORG

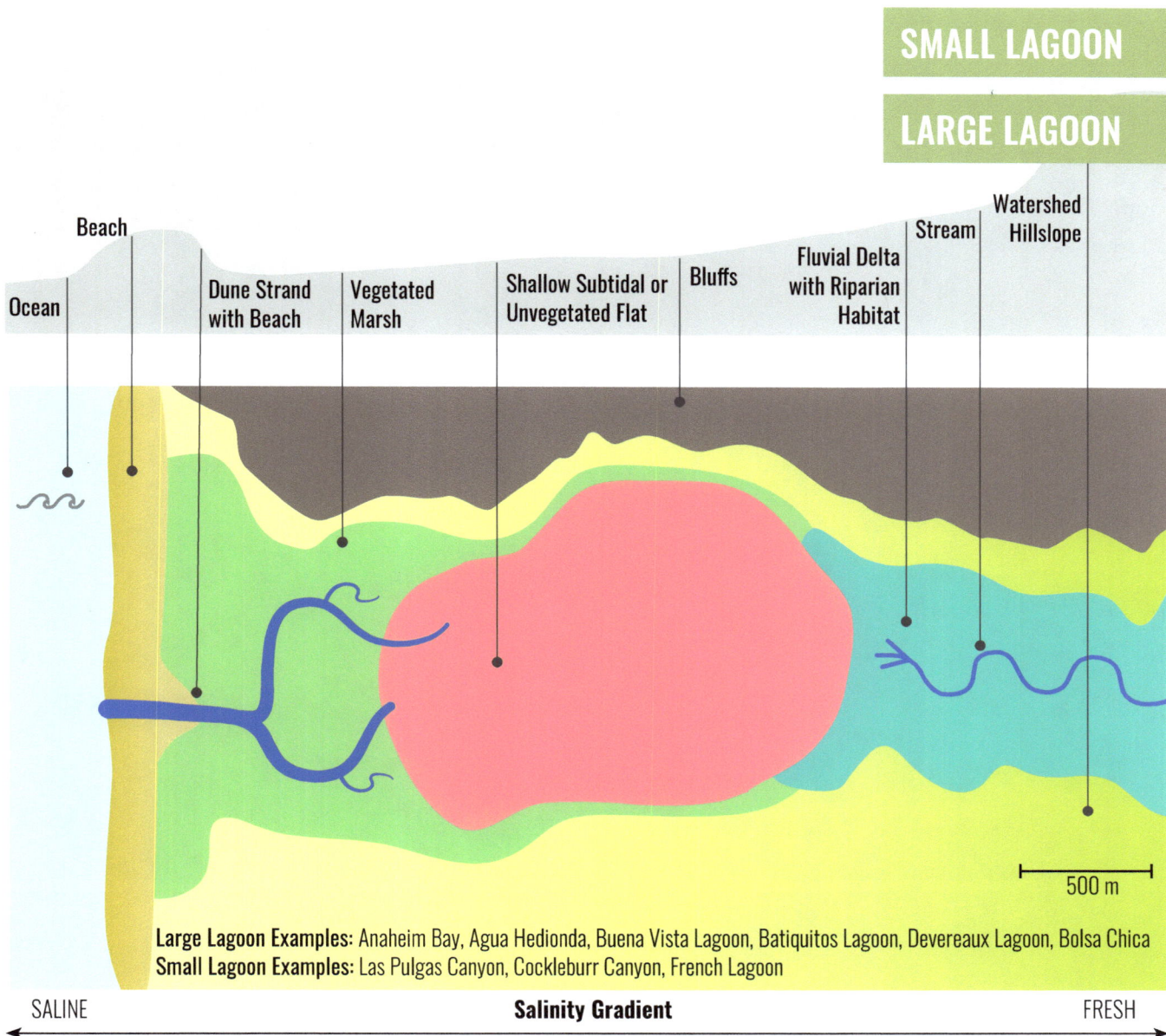

Ocean   Beach   Dune Strand with Beach   Vegetated Marsh   Shallow Subtidal or Unvegetated Flat   Bluffs   Fluvial Delta with Riparian Habitat   Stream   Watershed Hillslope

500 m

Large Lagoon Examples: Anaheim Bay, Agua Hedionda, Buena Vista Lagoon, Batiquitos Lagoon, Devereaux Lagoon, Bolsa Chica
Small Lagoon Examples: Las Pulgas Canyon, Cockleburr Canyon, French Lagoon

SALINE   **Salinity Gradient**   FRESH

Figure 7. Diagram of the small lagoon and large lagoon archetypes.

Definition of large lagoon and small lagoon: Large and small lagoons are shallow basins usually created by a beach berm or barrier, which traps the lagoon between the ocean and uplands. Historically, many of these had wide, flat basins with extensive unvegetated flats. These flats were intermittently-flooded on a seasonal or longer basis as the inlet opened or closed. Depending on the water level when the inlet closed and the length of closure, areas of ponded water may have dried completely to become salt flats. The large lagoons have larger tidal prisms than the smaller lagoons but not necessarily a larger watershed; any river flow may be relatively small and intermittent resulting in

more frequent closure of smaller lagoons. Sedimentation from the watershed may be limited compared to the area of the lagoon and may be deposited as an alluvial fan within the lagoon. Oceanic sediment may be deposited as a flood delta in the inlet. These lagoons may, therefore, have small intertidal areas, and may only have fringing vegetated marsh habitat. Today many of these lagoons have been modified, particularly by stabilizing the inlets and dredging them to be deeper. These management actions have created more consistent, less natural conditions within the lagoons, with the unvegetated flats now becoming more permanently flooded. The surrounding vegetated marsh habitats now have regular tidal inundation. See next page (top) for an example of a large lagoon (Batiquitos Lagoon).

EXAMPLE OF LARGE LAGOON: BATIQUITOS LAGOON • PHOTO BY TIM BUSS, COURTESY OF CREATIVE COMMONS

EXAMPLE OF INTERMEDIATE ESTUARY: DUME LAGOON • COPYRIGHT (C) 2002-2018 K. AND G. ADELMAN, CALIFORNIA COASTAL RECORDS PROJECT, WWW.CALIFORNIACOASTLINE.ORG

Ocean

Beach

Dune Strand
with Beach

Vegetated
Marsh

River

Riparian Area and
Adjacent Habitat

Examples: Malibu Lagoon, San Mateo Lagoon, Carpinteria Salt Marsh, Mugu Lagoon

500 m

SALINE

Salinity Gradient

FRESH

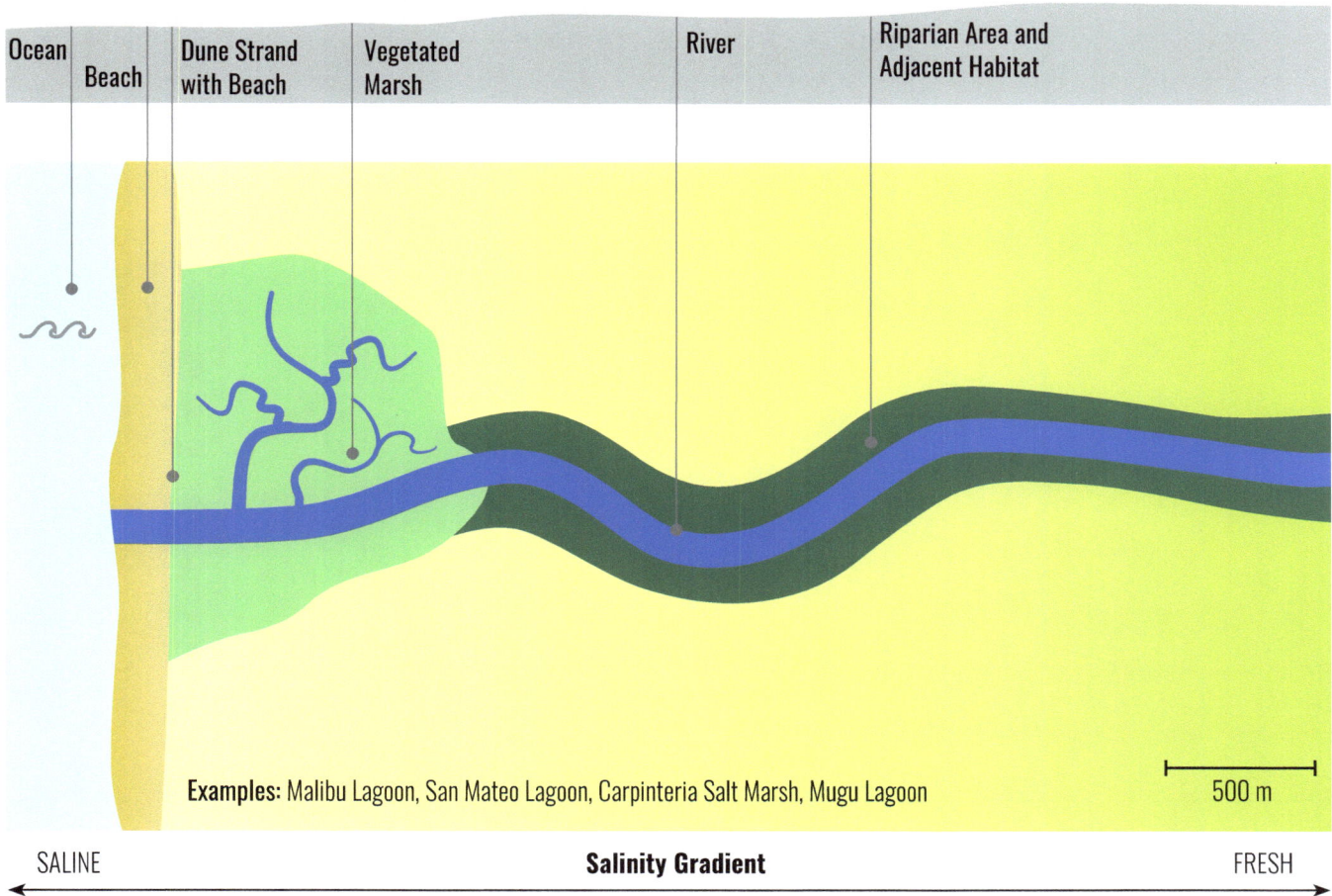

Figure 8. Diagram of the intermediate estuary archetype.

Definition of intermediate estuary: Intermediate estuaries lie between the large and small systems and have significant tidal prism and river flows. When they are closed, water levels within these estuaries are affected by river flow, if present, runoff from the immediate watershed, waves that overtop the berm, tides which affect groundwater elevations, seepage through the berm from the ocean, evapotranspiration, and overtopping on extreme tides. All of these processes are likely to affect water levels within the estuary and affect the likelihood and duration of opening, perching, or closing. Another controlling factor, tidal prism, similarly affects water levels and the probability and duration of inlet opening.

LARGE RIVER VALLEY ESTUARY

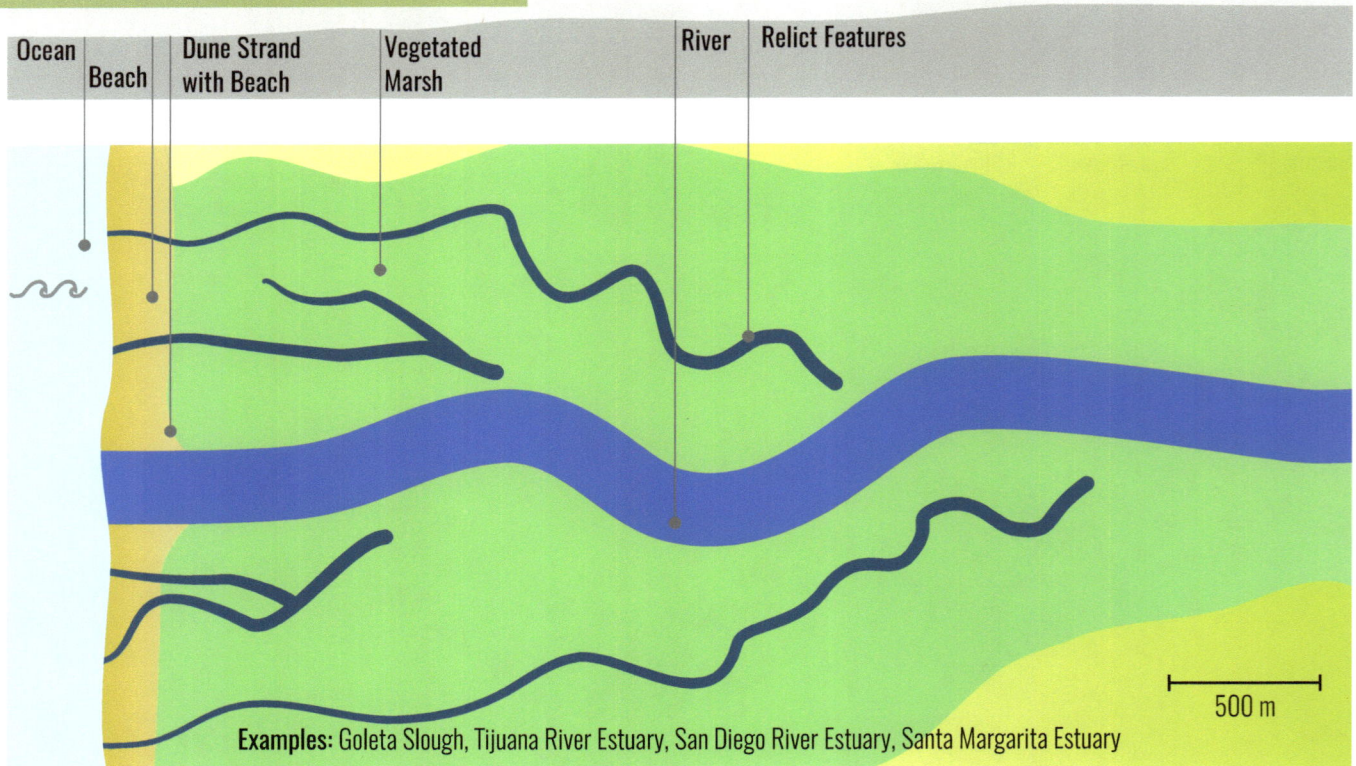

Ocean | Beach | Dune Strand with Beach | Vegetated Marsh | River | Relict Features

500 m

Examples: Goleta Slough, Tijuana River Estuary, San Diego River Estuary, Santa Margarita Estuary

FRAGMENTED RIVER VALLEY ESTUARY

SALINE | **Salinity Gradient** | FRESH

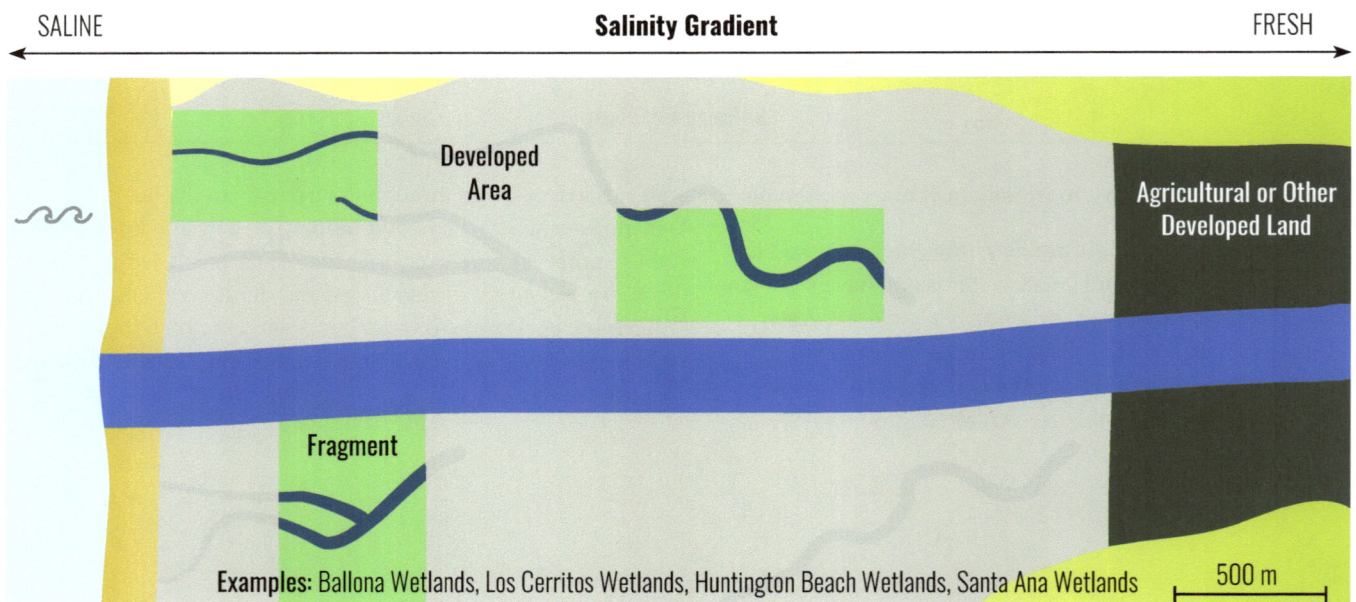

Developed Area

Agricultural or Other Developed Land

Fragment

500 m

Examples: Ballona Wetlands, Los Cerritos Wetlands, Huntington Beach Wetlands, Santa Ana Wetlands

Figure 9. Diagram depicting the large river valley estuary and the fragmented river valley estuary archetypes.

Definition of large river valley estuary and fragmented river valley estuary: These large, relatively flat and easily drained plains have been very attractive for development. As a result, they have been drained, diked, and developed, fragmenting the floodplain and wetlands. Some river channels have been completely rerouted to facilitate this drainage and to improve flood protection. This has led to the fragmentation of the large river valley estuaries where remnants of the floodplain have been dissected into smaller, spatially distinct units. The habitats within these fragments will not necessarily reflect the diversity or proportions of habitats of the undisturbed wetlands units. Even where the wetlands remain connected, larger rivers that fed these wetlands have tended to be dammed—trapping water and sediment.

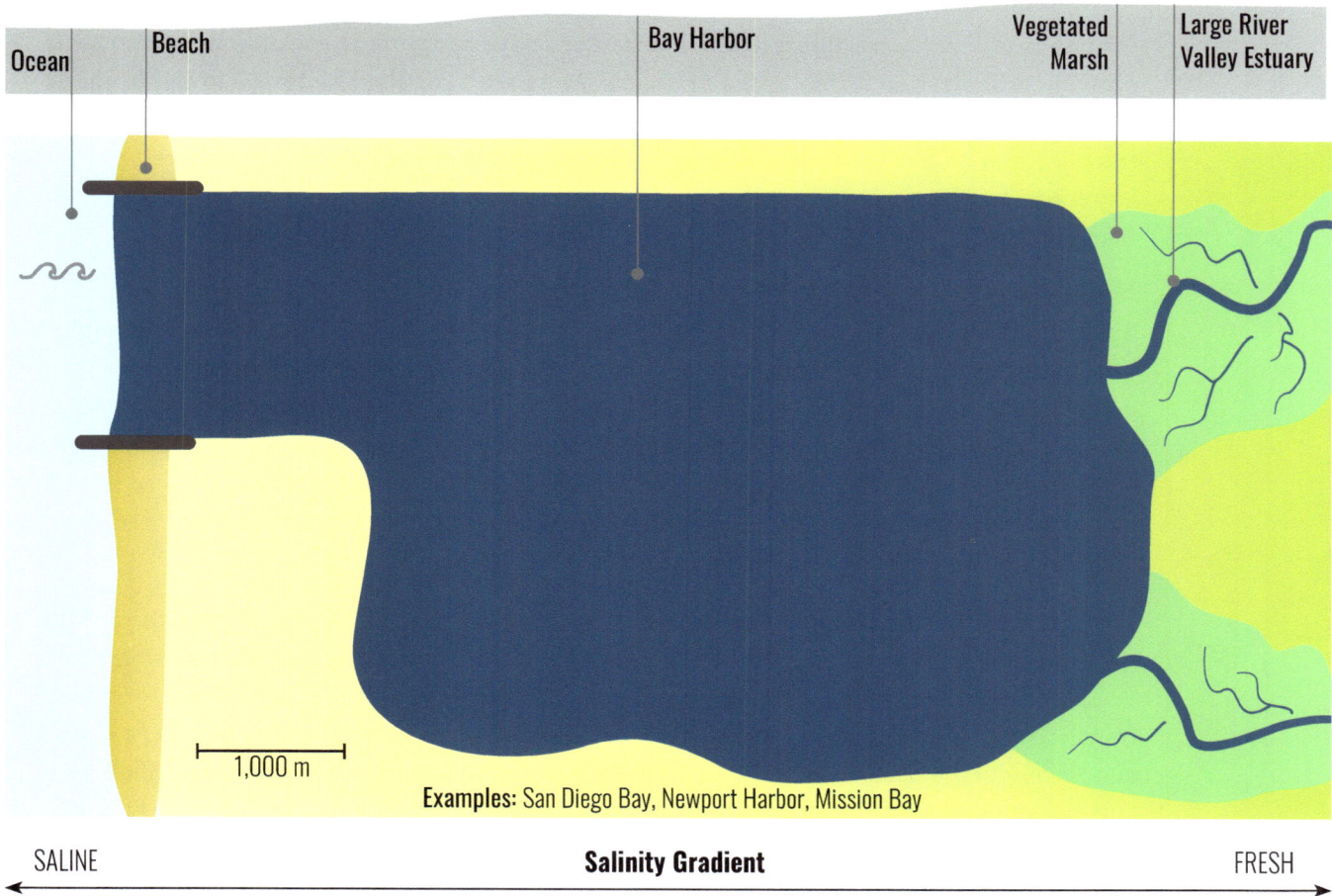

Ocean | Beach | Bay Harbor | Vegetated Marsh | Large River Valley Estuary

1,000 m

Examples: San Diego Bay, Newport Harbor, Mission Bay

SALINE | **Salinity Gradient** | FRESH

Figure 10. Diagram of the open bay/harbor archetype.

Definition of open bay/harbor: Open bays and harbors are tidally-dominated, have large tidal prisms, small river inputs, significant subtidal areas, relatively little intertidal wetlands and permanently open inlets. Many have hardened mouth infrastructure to help maintain tidal action, reduce sedimentation, and provide for safe harbor usage. These archetypes are relatively large compared to their sediment supply and have not filled in.

EXAMPLE OF OPEN HARBOR: NEWPORT HARBOR • PHOTO BY CHRIS JEPSON, COURTESY OF CREATIVE COMMONS

COASTAL WETLAND CHANGES OVER TIME

Wetlands in Southern California are under intense stress from the region's significant and growing urbanization. While some of the stressors are a result of historical impacts, many impacts to wetlands are ongoing and some, like climate change, are predicted to occur in the future. The following sections discuss the stressors and resulting changes to wetland form and function that create the conditions in which wetlands preservation and restoration must take place.

Historical Losses and Current Stressors

Southern California's wetlands historically supported large areas of vegetated and unvegetated tidal marsh. Around 1850 there were approximately 33,400 acres (13,510 ha) of all coastal wetland habitats (subtidal, unvegetated flats and vegetated marsh)—not counting 14,948 acres (6,050 ha) of subtidal embayments such as San Diego Bay (Figure 11A). In the following 150 years, more than 62% of these wetlands have been lost. By 2005 only 12,800 acres (5,180 ha) remained (Figure 11B). Vegetated marsh has experienced the greatest absolute decline with a loss of 13,400 acres (5,420 ha) (Figure 11B).

The historical loss and fragmentation of coastal wetlands that has occurred since the 1800s has resulted from the intensive urban development that characterizes Southern California. Wetlands have been diked, drained, and broken into pieces to allow for agriculture, grazing, transportation, development, and flood protection. For many tidal wetlands, the hydrologic connection to the ocean has also been modified as tidal inlets have been dredged, filled and trained to maintain open tidal connections, reduce inlet migration, and address inland flooding. These many wetland stressors have reduced large wetland complexes down to small fragmented systems.

Hydromodification upstream within watersheds has also impacted coastal wetlands through modifications such as dam construction, water diversion, and urban runoff. Hydromodification has drastically altered the wetland composition of watersheds due to changes in the timing and composition of water and sediment flow. Historically, with a Mediterranean climate, the streamflows in Southern California were highly seasonal, with the bulk of freshwater and sediment flow occurring during the wet season and little surface flow reaching the coastal wetlands during much of the dry season (Beller at al 2014). However, early watershed modifications, such as dam construction and surface and groundwater diversions, likely decreased freshwater and sediment inputs to the watersheds. In contrast, urban runoff, irrigation, and wastewater discharge associated with urban development in the mid- to late 20th century has tended to increase freshwater and sediment inputs. As a result, urbanized wetland systems receive greater concentrations of freshwater and sediment, and greater peak water flows. Year-round urban runoff can, at times, result in a year-round supply of water and sediment in coastal wetlands. These changes in water and sediment flow have resulted in shifts in sediment elevation, habitat composition, and species usage.

These modifications within wetlands and their associated watersheds have not only reduced wetland areas, but the composition of wetland habitat has also changed. For instance Stein et al. (2014) reported a 43% conversion of wetlands to uplands and developed areas. In addition to a decrease in the area of unvegetated flats and vegetated marsh, subtidal areas have increased in area by 100%, mainly due to dredging of marinas and harbors (Figures 11A and 11B).

Degraded wetlands that are not functioning under natural conditions are more vulnerable to non-native species invasions, further facilitating habitat conversion (Werner and Zedler 2002). These non-native species can cause changes in habitat composition and ecosystem function. For instance, invasive vegetation in coastal wetlands can form monotypic stands changing the habitat structure, lowering biodiversity, altering nutrient cycling, and cascading changes throughout the food web (Zedler and Kercher 2004). With Southern California's history of subjecting wetlands to numerous stressors, the region's coastal wetlands are in turn particularly sensitive to invasion.

Historical wetland loss has varied by subregion (Figure 13). The San Pedro Bay and San Diego Coast subregions historically accounted for more than 80% of all historical coastal wetland habitats. Today, wetland habitat in the San Pedro Bay subregion has been reduced by 93% of its historical extent, mostly from the loss of large river valley estuaries. The Santa Monica Bay subregion experienced a 91% loss, while the Ventura coast shows the least wetland loss, with a reduction of about 30% of vegetated and unvegetated habitats.

Future Wetland Losses

In addition to the wetland losses and conversion that occurred in the past, more wetlands will be lost in the future to sea-level rise. Higher sea levels will lead to changes in coastal wetlands as we know them; remaining wetlands will be squeezed between the ocean and the developed land, leaving little space for existing wetlands to survive. WRP agencies have invested billions of dollars in coastal wetlands restoration—that investment would be lost largely due to sea-level rise. We have opportunities to prevent this loss starting today by acquiring lands adjacent to coastal wetland and immediately starting to facilitate wetland migration into those areas.

The WRP (through an effort lead by the Southern California Coastal Water Research Project) developed a model to predict wetland habitat change resulting from sea-level rise. This habitat change model estimates how much coastal wetland area could be lost under two sea-level rise scenarios, and predicts the amount of habitat change in different wetland archetypes. The model also identifies opportunities for wetland migration, which managers can use to plan for future expansion of tidal wetlands in response to rising sea level. Wetland migration modeling was conducted in three possible footprints: no migration, migration avoiding developed areas, and migration including developed areas (Figure 12). The areas that will be available for wetlands in the future are today's wetland-upland transition zones (transition zones), which exist at the landward edge of the wetlands. For detailed information on how this sea-level rise analysis was completed, see Appendix 3.

According to the habitat change model, approximately 800 acres (320 ha) of vegetated marsh and unvegetated flats will be lost region-wide under the 24-inch sea-level rise scenario, and 3,700 acres (1,500 ha) will be lost under the 66-inch sea-level rise scenario (Figure 11C). These predicted losses assume that no action will be taken to allow for wetland migration into current upland habitat.

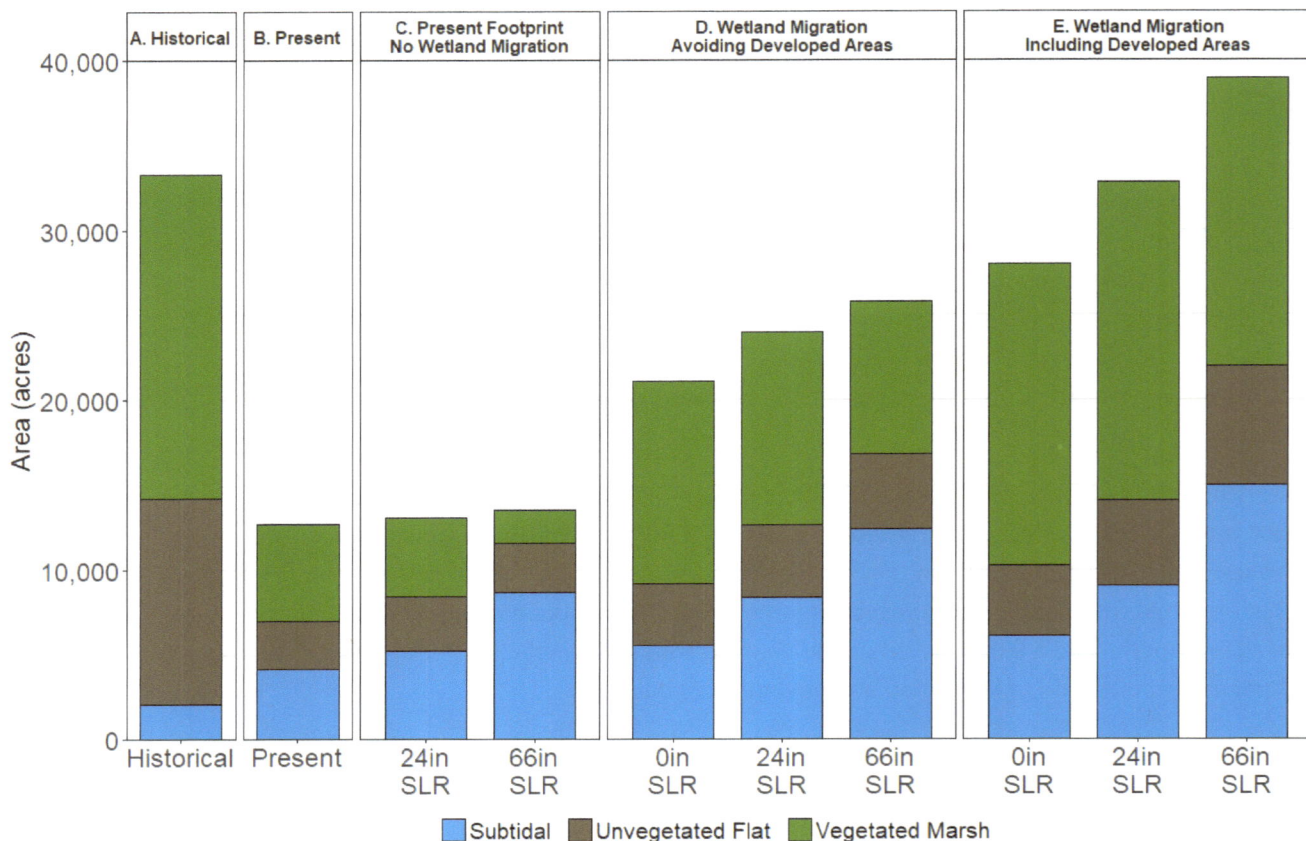

Figures 11A–11E. Change in overall extent and composition of coastal wetlands between (A) historical, (B) present and (D & E) future habitat distribution (excluding San Diego Bay subtidal). The present/future distribution is based on three sea-level rise scenarios (current, 24 inches, and 66 inches) and three wetland boundaries (C) no wetland migration/existing wetland footprint, (D) wetland migration/avoid developed areas, and (E) wetland migration/all areas.

Figure 12. Map visualization of the footprints used in the analyses for each wetland migration scenario: (A) no wetland migration, (B) wetland migration avoiding developed areas and (C) wetland migration including developed areas in Los Peñasquitos.

MODELING CHANGES TO TIDAL INLETS

Future rates of tidal inlet opening and closure, and resultant water level changes, were determined using a simplified water balance model with inputs from regional datasets (Appendix 1). The simplified model demonstrated a likelihood for tidal inlets to close more frequently with higher sea levels. The magnitude of the increase in the rate of closures is uncertain due to the use of regional data and the number of assumptions that were made in running the simplified model. The WRP will expand upon this modeling work with a new project funded by the National Oceanic and Atmospheric Administration (NOAA). This NOAA-funded project will estimate future rates of tidal inlet closures, the subsequent changes in water level and salinity, and how those changes will affect coastal wetland habitats in Southern California.

Wetland losses from projected sea-level rise could be offset by facilitating migration of wetlands into adjacent uplands. Our modelling predicts that after 24 inches of sea-level rise, 7,700 acres of current upland habitat could become vegetated marsh and flats if wetlands were able to expand into undeveloped areas (wetland migration/avoid developed areas) (Figure 11D). Even larger areas will be needed to accommodate wetland migration with 66 inches of sea-level rise. Under this scenario, 8,700 acres (3,520 ha) of current upland habitat could become vegetated marsh and flats (Figure 11D).

Restoration Opportunities

Despite the historical and projected losses, opportunities exist for wetlands to migrate into adjacent undeveloped lands, which would require realignment of physical barriers such as levees, roads, and other infrastructure (Figure 11D). Even more opportunities for wetlands to migrate into adjacent land would exist if infrastructure and development were removed (Figure 11E). These wetland restoration opportunities vary regionally (Figure 13). The greatest opportunities to facilitate wetlands restoration in the future lie in the Ventura, San Pedro, and San Diego subregions, because these subregions have larger wetlands. For example, large lagoons provide greater opportunities for wetland migration into adjacent transition zones (Appendix 3). Santa Barbara and Santa Monica Bay also have opportunities, but these regions generally have smaller wetland systems. Where there is significantly less urbanization, such as the Ventura coast, there is more restorable agricultural land than in the more heavily urbanized coastal places like San Pedro Bay. These less-developed areas are where restoration actions could most efficiently facilitate the migration of wetland habitat and lessen the future impacts of sea-level rise.

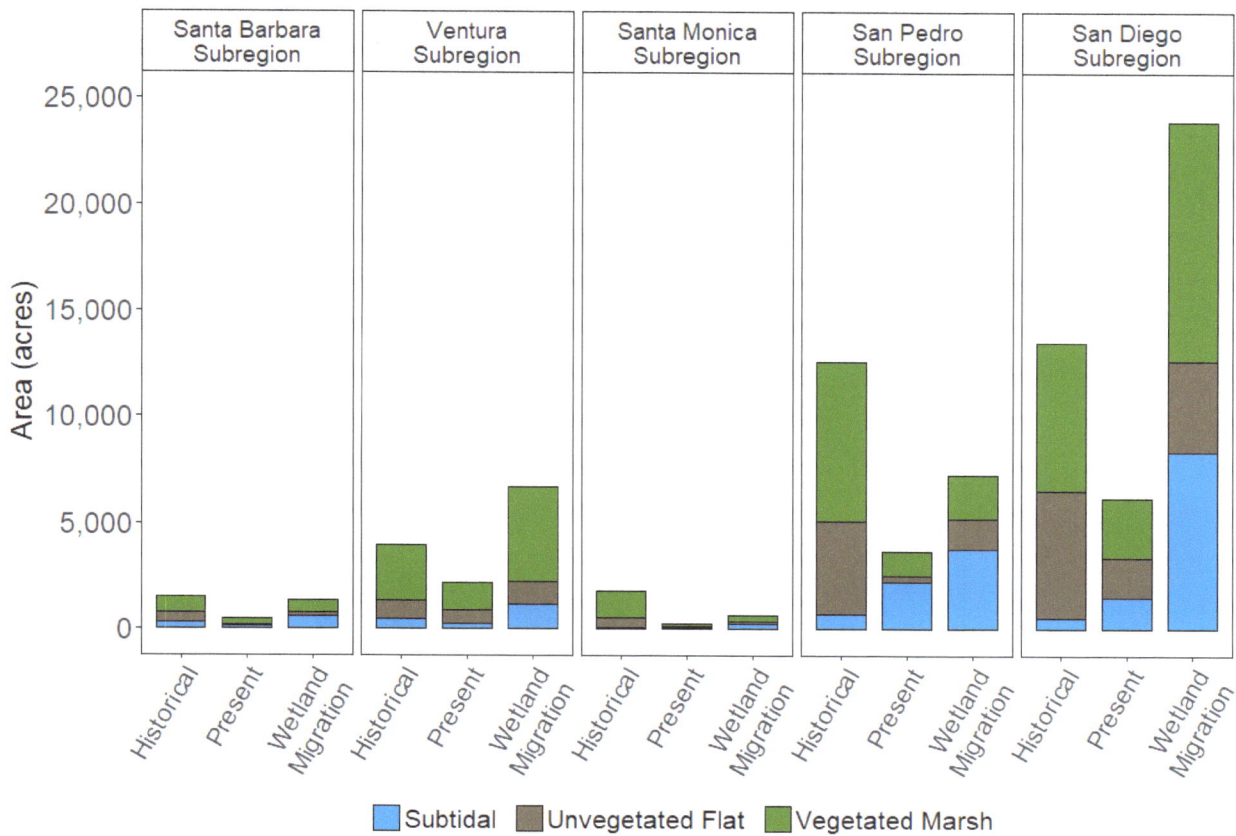

Figure 13. Proportion of historical, present, and future coastal wetland habitat types by subregion. The future/wetland migration distribution is based on a 24-inch sea-level rise scenario, excluding migration into developed areas.

Conclusion

Sea-level rise both threatens our past economic, ecologic, and social investment in wetlands protection and restoration, and calls on us to make even greater investments going forward to preserve our wetlands into the future. Higher sea levels will result in wetland loss and habitat type conversion, which will threaten our ability to maintain a network of diverse wetlands along the Southern California coast. However, the worst losses will not occur until sea-level rise exceeds 24 inches (currently projected around 2050). This provides time for deliberate action to acquire uplands adjacent to current wetlands, to implement wetland restoration, and to facilitate wetland management in ways that can better accommodate projected sea-level rise. The following chapters provide Goals and Objectives that provide a roadmap for increasing the resilience of Southern California's network of coastal wetlands to sea-level rise.

GOAL 1: RESTORE COASTAL WETLANDS
OBJECTIVES AND MANAGEMENT STRATEGIES

The following seven Objectives provide the quantitative targets for the WRP's partners to help accomplish Goal 1, *preserve and restore resilient coastal tidal wetlands and associated marine and terrestrial habitats*. These objectives are based on the understanding of historical, current and predicted likely future wetland distribution developed through the analyses presented in the previous pages. The Objectives for Goal 1 address the abundance of coastal wetlands, their diversity, connectivity, and condition, to restore functions and processes and promote resilience to climate change and other stressors. Along with a set of Management Strategies, the seven Objectives are intended to help ensure that WRP projects support achievement of Goal 1 by providing a regional perspective and a method to quantify progress. The Objectives were developed to be aspirational; not all Objectives are suitable and achievable for every coastal wetland in Southern California.

Implementation of the Objectives would include the continued management of existing wetlands, the restoration of potential restorable areas today, and the protection and enhancement of wetland-upland transition zone that may become tomorrow's wetlands as sea levels rise. Accomplishing these coastal wetland Objectives will likely require management of the estuary-watershed system (i.e., managing water, sediment, and constituent inputs from the watershed, as well as inputs from the ocean). This may result in considerations of retrofitting, removing or modifying existing structures and management approaches.

There will always be costs and benefits associated with wetlands restoration, and decisions on how and where to implement restoration will only get more complicated as sea levels rise. Benefits of diking wetlands, such as corridors for transportation and land for housing must be weighed against benefits of restoration such as public access to open space, resilience to sea-level rise, increased carbon sequestration, and wildlife protection. In order to truly compare the costs and benefits of wetlands restoration, an economic valuation of wetland benefits would be needed, which was not part of the scope of this effort. But even without a cost-benefit analysis, we know that wetlands have been undervalued in the past (Ballard et al. 2016), and the pressures and impacts on these systems are ever-increasing. Without an aspirational approach for the region we may fall significantly short of our goals. The costs and benefits of each wetlands restoration project should be analyzed at the site-specific level in relation to the objectives of the project.

A quick glance at the quantitative Objectives for the recovery of coastal wetlands (Goal 1) is presented in Table 1. Each of these Objectives, and associated Management Strategies, are described in detail in the sections that follow.

Table 1. A **summary of the seven Objectives** that comprise Goal 1 of the *Regional Strategy 2018*.

Objective	Description	Management Strategies
1. Restore Wetland Area	A. Preserve 8,600 acres (3,480 ha) of existing wetlands. B. Facilitate wetland migration and restoration of 7,700 acres (3,116 ha) after 24 inches of sea-level rise. C. This restoration and facilitation will result in 15,500 acres (6,273 ha) of wetland habitat after 24 inches of sea-level rise	1. Remove barriers that prevent wetlands from expanding or migrating. 2. Protect, manage and acquire adjacent land. 3. Grade areas adjacent to wetlands to increase opportunity for migration. 4. Relocate or modify adjacent infrastructure or development.
2. Restore Wetland Size	Increase coastal wetland size in areas where 24 inches of sea-level rise will support wetlands in the future, to more closely approximate historical distribution within each subregion.	Same as 1–4
3. Restore Wetland Archetype Distribution	A. Preserve or restore, as appropriate, the historical distribution of archetypes in each subregion. B. Increase and maintain connectivity between historically connected wetland fragments.	5. Remove barriers to reconnect channels to wetlands. 6. Allow tidal inlets to open and close naturally. 7. Modify or remove structures to restore inundation regime.
4. Habitat Diversity	Restore or maintain the coastal wetland habitat composition, represented by the historical archetype habitat profiles, in at least 50% of the systems within a given archetype across a subregion.	Same as 5–7, and: 8. Protect existing natural salt flats and their supporting hydrological regime, while also protecting anthropogenic salt flats where it can be demonstrated they have value that other habitats within the system cannot support. 9. Protect existing shallow subtidal habitats associated with coastal wetlands.
5. Wetland-Upland Transition Zone	A. Protect all existing natural areas of wetland-upland transition zones up to 1,600 feet (500m) from the marsh edge. New structures within transition zones should be minimal, not impede wetland migration, and potentially removable. B. Increase area of natural wetland-upland transition zone to facilitate marsh migration, so that at least a minimum of 40% of the wetland perimeter is bounded by transition zone that extends inland for at least the full estimated tidal extent under 24 inches (0.6 m) of sea level rise. C. Increase areas of natural wetland-upland transition zone up to 1,600 feet (500m) from the marsh edge, even in areas that are not contiguous with the marsh. D. If the system has a river or creek , then an additional focus should be the creation of adjacent habitat that allows for the upstream migration of wetlands, at least to the head of tide under 24 inches (0.6 m) of sea level rise.	Same as 1–4, and: 10. Protect, manage and acquire adjacent land within the wetland-upland transition zone.

Table 1 (continued). A summary of the seven Objectives that comprise Goal 1 of the *Regional Strategy 2018.*

Objective	Description	Management Strategies
6. Restore Hydrological Connectivity	A. Restore tidal characteristics (range, extent and residence time), guided by appropriate reference conditions, to support habitat abundance and distribution as indicated in Objectives 1–4. B. Restore freshwater and sediment flow characteristics from watersheds (volume, frequency, and timing), guided by appropriate reference conditions, to support habitat abundance and distribution as indicated in Objectives 1–4. C. Restore or manage sediment inputs to maintain wetland and wetland-upland transition zone elevations sufficient to accommodate 24 inches (0.6 m) of estimated sea level rise. Inputs should be assessed based on total annual volume and magnitude of peak inputs.	Same as 5–7, and: 11. Remove barriers to release sediment held higher in the watershed. 12. Manage flows in river channels to increase their capacity to move sediment from the watershed. 13. Augment sediment processes to raise and maintain marsh elevation.
7. Wetland Condition	A. Improve the major attributes of wetland condition, including biology, hydrology, physical structure, and landscape context, as measured by a rapid assessment score, for 100% of systems within each archetype. B. 100% of mature coastal wetlands (i.e., natural coastal wetland or restored coastal wetland of 40 years or more) should achieve and maintain an overall CRAM score ranging from 76–94. C. 100% of future restoration projects should be on or above the Habitat Development Curve based on the project age as the restoration matures.	14. Review pre-construction CRAM score to determine what needs to be restored in project design in order to make a CRAM score of 76–94. 15. Review post-construction CRAM score and compare to project's evolution to the Habitat Development Curve.

A detailed description of each Objective can be found below. The following structure is used for each Objective:

- Objective;

- Rationale for the Objective;

- Management strategies that will help accomplish the Objective; and

- Recommended methodology to track the Objective.

Objectives that apply equally to all the subregions are summarized at the regional scale; specific Objectives for individual subregions are detailed as appropriate. For more information on the data types used to develop each Objective, see Appendix 4.

A. Preserve the 8,600 acres of existing wetlands.

B. Facilitate wetland migration and restoration of an additional 7,700 acres (3,110 ha) of wetlands after 24 inches of sea-level rise. Achieving 7,700 acres (3,110 ha) of wetlands after 24 inches (0.6 m) of sea-level rise would require restoration of both today's higher-elevation wetlands and facilitation of wetland migration (i.e., transition zone/upland habitat).

C. Preserving and restoring wetlands today and facilitating wetland migration could result in a total of 15,500 acres (6,270 ha) of wetlands by the time sea-level rise reaches 24 inches (0.6 m). Achieving 15,500 (6,270 ha) acres of coastal wetland area will require the acquisition of approximately 7,400 acres (3,000 ha) of private land, both presently diked and drained, and of adjacent upland areas, as indicated in Table 2 below.

Table 2. Present and future coastal wetland area by subregion. The rounded total is how these Objectives are represented throughout the rest of the document.

Subregion	Tidal Flats and Marshes Acres (Hectares)				
	Preserve Existing Wetlands	**Expected Net Losses/ Gains With 24 inches of Sea-Level Rise Within the Existing Wetland Footprint**	**Restore and Facilitate Wetland Migration (with 24 inches of sea-level rise)**	**Total Wetlands with 24 inches of Sea-Level Rise**	**Private Land to Acquire (out of "Total")**
Santa Barbara	401 (162)	-91 (-37)	423 (171)	733 (296)	263 (106)
Ventura	1,911 (773)	75 (31)	3,535 (1,431)	5,521 (2,235)	4,029 (1,630)
Santa Monica	186 (75)	1 (0)	234 (95)	421 (170)	36 (15)
San Pedro	1,431 (579)	-12 (-5)	2,036 (824)	3,455 (1,398)	1,493 (604)
San Diego	4,686 (1,896)	-791 (-320)	1,526 (618)	5,421 (2,194)	1,660 (672)
Total*	**8,615 (3,486)**	**-818 (-331)**	**7,754 (3,139)**	**15,551 (6,293)**	**7,482 (3,028)**
Total Rounded	**8,600 (3,480)**	**-800 (-324)**	**7,700 (3,116)**	**15,500 (6,273)**	**7,400 (2,995)**

* excludes all present day harbors, ports, and marinas

Rationale: Extensive areas of wetlands have been lost over the last 200 years (pages 24–25), and those losses will only be exacerbated by rising sea levels (pages 25–27). The historical, existing and future coastal wetland areas for the region are shown in Figure 14. Even if we are able to facilitate wetland restoration in all of the areas that are uplands today but will be at wetland elevation in the future with sea level rise, we will not regain the historical wetlands that have been lost.

HISTORICAL AND FUTURE WETLANDS

The full implementation of this Objective will partially recover the historical wetland area that has been lost. However, these wetlands will not be in their historical locations because many historical wetland areas will never be restorable. If coastal wetlands are going to persist in the future, the WRP's partners must take action to facilitate the restoration of additional wetland areas. The majority of the potential wetland restoration areas will be in the wetland-upland transition zones because sea-level rise will convert upland areas to wetlands.

The quantitative Objective to preserve the 8,600 acres (3,480 ha) of existing coastal wetland habitat was calculated using maps of current wetland extent, assuming that all existing wetland areas will be protected. It is estimated that up to 800 acres (320 ha) of these existing coastal wetlands may be lost with 24 inches (0.6 m) of sea-level rise.

The potential restoration area of 7,700 acres (3,110 ha) includes currently undeveloped drained and diked lands, as well as adjacent uplands that could be restored to higher-elevation wetland today or could be tidally inundated after 24 inches (0.6 m) of sea-level rise if required activities to restore hydrologic connectivity are implemented (page 66). This restoration Objective focuses on restoring higher elevation wetland areas, as well as allowing areas that are currently wetland-upland transition zones to become wetlands in the future.

There are many areas available today for tidal wetland restoration, which include undeveloped areas and some developed areas like parking lots and agriculture. However, some of these areas may not be the most resilient to future sea level rise. Some of that area, once restored, would result in low and mid-elevation wetland habitats that are more vulnerable to habitat conversion in the face of sea-level rise. Due to this concern, Objective 1 focuses on restoring the higher-elevation wetland and transition zones in order to achieve 7,700 acres of additional restored wetland habitat at 24 inches of sea level rise.

Of the total 15,500 acres of wetlands to protect and restore, approximately 7,400 acres are currently privately-held. Some of this land is currently wetland and some of it is upland that will become wetland with 24 inches of sea-level rise.

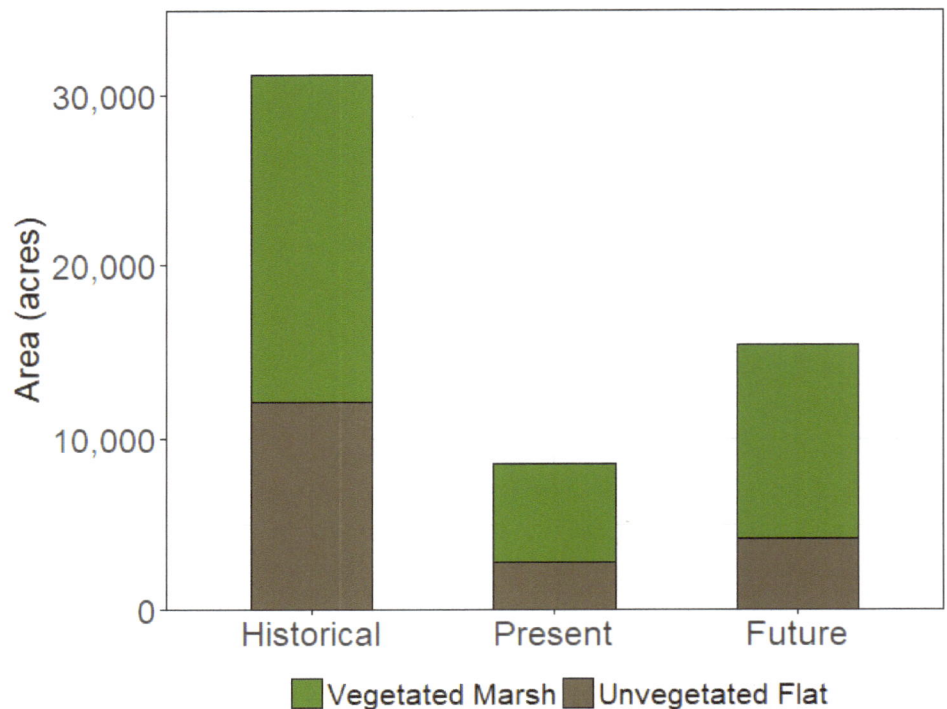

Figure 14. Historical, present, and restored future coastal wetland area. The future area represents the wetland area after active restoration (e.g., reconnecting lands to tidal action) and 24 inches (60.8 centimeters) of sea-level rise.

Management Strategies: To achieve 15,500 acres (6,270 ha) of healthy, resilient coastal wetlands and reduce sea-level-rise induced habitat loss, the WRP recommends four key Management Strategies (Figure 15). These four Management Strategies focus on allowing wetlands to expand and migrate by changing the landscape to facilitate migration and/or reconnect currently fragmented wetlands. As demonstrated in Appendix 3, wetland migration is the most effective strategy for maintaining wetland habitat in the face of sea-level rise. One key action that will facilitate wetland migration is the acquisition and protection of transition zones. Transition zones offer immediate and future value by providing an opportunity to establish habitats that have been lost from the existing landscape, and also to establish areas for wetlands to migrate in the future.

In order for wetlands to migrate into adjacent upland transition zones, barriers that prevent wetland migration must be removed and suitable upland habitat conditions must be created. In many cases, upland transition zones are confined by existing flood risk management berms. The areas outside existing berms could be transition zone habitat if barriers were removed. In the case of rivers and streams, transition zones will need to be restored upstream. Since transitions zones are typically outside existing wetland boundaries, acquisition and protection of adjacent open space should be the utmost priority. Available transition zones vary by system and archetype, but all opportunities to acquire and protect adjacent or upstream transition zones should be pursued.

For small creeks and lagoons where wetland migration is limited by geography and topography, removal of barriers to inland and upstream migration is a high priority. Considering most of these wetlands will be trapped due to steep topographies (either natural, like steep valley sides, or levees), man-made barriers that prevent migration should be removed to avoid vegetated marsh loss. However, many small creeks and lagoons will not be able to migrate with sea-level rise and management actions for these trapped systems will include learning to appreciate the ecosystem values of the fringing marshes and subtidal habitats that may remain.

While not every Management Strategy will work for every site or situation, they should be applied based on current constraints and opportunities, political and community support, and funding availability.

> **Management Strategy 1: Remove barriers that prevent wetlands from expanding or migrating.** Many wetlands are surrounded by berms and levees to protect adjacent areas from flooding. These barriers may be removed or lowered to allow wetlands to migrate to keep pace with sea-level rise. Complete removal of barriers is preferred, as opposed to simply reconnecting water flow (e.g., via a culvert or pipe), in order to facilitate successful wetland migration.

> **Management Strategy 2: Protect, manage, and acquire adjacent land within the wetland migration zone.** Areas that may be suitable as wetland migration zones are not necessarily in public ownership and may be subject to development pressures, making land acquisition in transition zones a challenge. Protecting adjacent open space either by acquisition or by easement should be a priority. In addition, areas protected or acquired for wetland migration will be vulnerable to invasive species if left unmanaged. Such land needs to be actively managed to successfully establish native species over invasive species.

> **Management Strategy 3: Grade areas adjacent to wetlands to increase opportunity for migration.** In some cases, the areas adjacent to wetlands have been filled or graded to different elevations and slopes, and it may be necessary to remove or add fill as needed. In addition, it is not always feasible to realign flood risk management berms, so artificial transition zones, shallower slopes on the seaward side of steep-sided berms (a.k.a. horizontal levee or ecotone slope), may need to be considered. This may

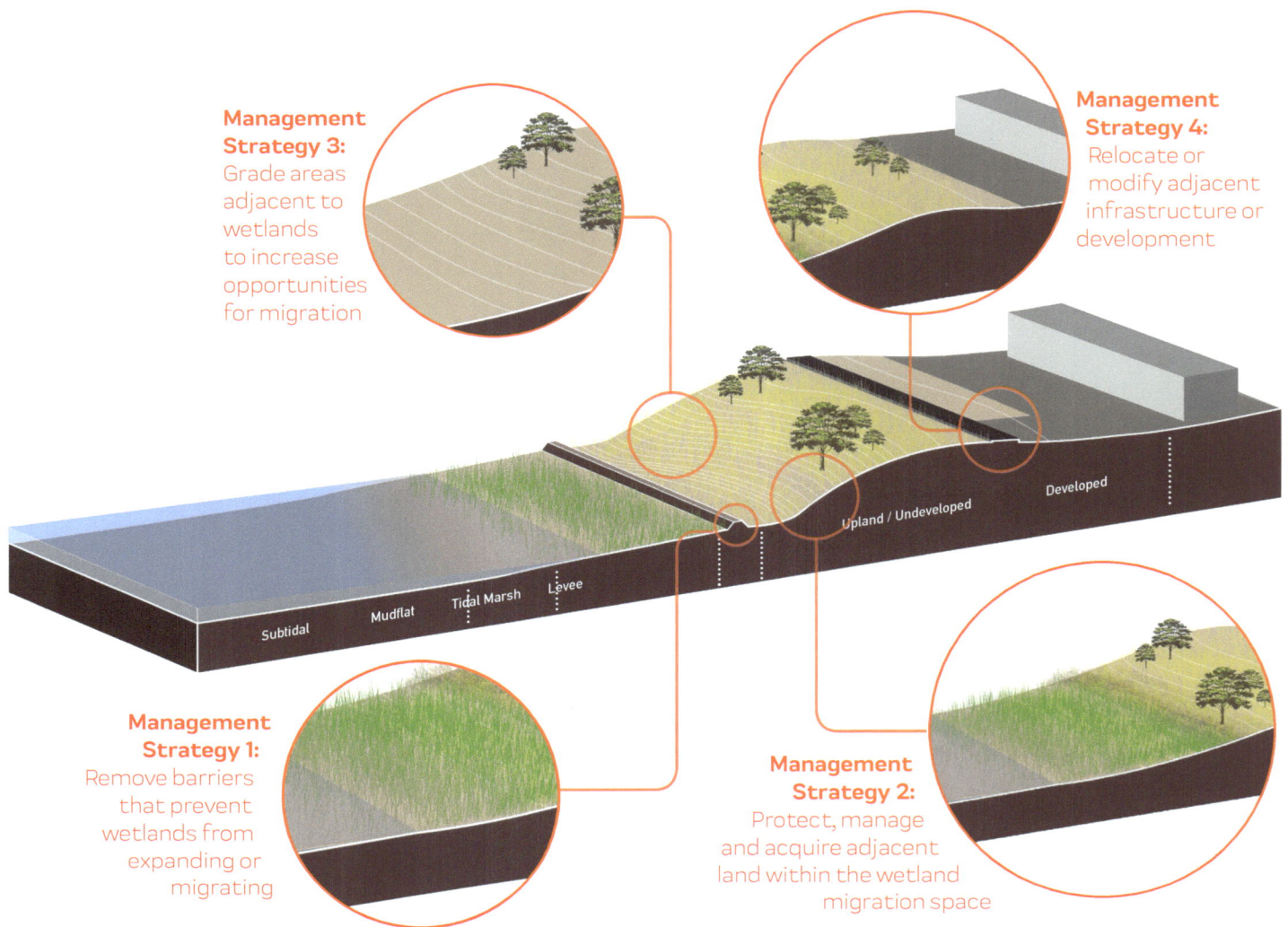

Management Strategy 3: Grade areas adjacent to wetlands to increase opportunities for migration

Management Strategy 4: Relocate or modify adjacent infrastructure or development

Management Strategy 1: Remove barriers that prevent wetlands from expanding or migrating

Management Strategy 2: Protect, manage and acquire adjacent land within the wetland migration space

Subtidal · Mudflat · Tidal Marsh · Levee · Upland / Undeveloped · Developed

Figure 15. Conceptual diagram of Management Strategies 1–4.

require a balance between short-term impacts, like the filling of existing wetlands if no available space exists landward of the berm, and long-term impacts.

Management Strategy 4: Relocate or modify adjacent infrastructure or development. Infrastructure is often present in transition zones (e.g., pipelines, transmission lines, trails, roads), which may decrease function in transition zone habitat, or which may be at risk of increased inundation due to sea-level rise. The removal and relocation of this infrastructure would allow for expansion of current wetland extents, and could be planned alongside infrastructure upgrades.

Objective Tracking: Measure the area of preserved wetlands, restored higher-elevation wetland, and facilitated wetland migration (i.e., transition zone/uplands today). This will be recorded on a per project basis and summarized for each subregion and region. This information will be derived from the complete habitat distribution map which will also be used to measure Objective 4: Restore Habitat Diversity.

Objective 2: Coastal Wetland Size

Increase coastal wetland size in areas where 24 inches (60.8 centimeters) of sea-level rise will support wetlands in the future, and to more closely approximate historical distribution within each subregion.

Rationale: The size of individual coastal wetlands is important to provide sufficient space to sustain key biological and physical processes and to support the redundancy, diversity and complexity, and connectivity necessary to foster resilience. Large areas provide room to accommodate landscape-scale processes and large, diverse populations. Larger wetlands correlate with greater species richness (Keddy et al. 2009), and are more resilient to disturbances (Moreno-Mateos et al. 2012). The average size of coastal wetlands in each subregion has decreased over time. The desire is to increase individual wetland sizes to approximate the historical size distributions, if there is available space to accommodate sea-level rise, in each subregion.

Figure 16 shows how the size of coastal wetlands in each of the subregions has changed over time and what size classes will have available space in the future. The light green bars labeled "Future" represent the 15,500-acre (6,270 ha) restoration in Objective 1, these are the areas that could be wetlands in the future with restoration action and facilitated wetland migration. This current Objective provides insight into the restoration size distribution. With the restoration of tidal action to leveed wetlands, there are opportunities for some systems to expand back towards their historical size.

Management Strategies: To achieve an increase in coastal wetland sizes, the WRP recommends the same Management Strategies (1–4; Figure 15) as for Objective 1. A majority of the opportunities to increase wetland sizes will be dependent on whether or not adjacent inland areas are available for inland wetland migration, similar to Objective 1. Facilitating wetland migration will be the most effective management strategy to achieve an overall increase in acreage within Southern California's urban landscape and in the face of sea-level rise.

Objective Tracking: Measure the area of individual coastal wetlands, including the restoration project area. Whereas monitoring for Objective 1 measures project size, monitoring for this Objective measures the size of the whole coastal wetland (e.g., subtidal, vegetated marsh, unvegetated flats, and wetland-upland transition zone). If a project increases the size class of the coastal wetland, takes advantage of available wetland space in the future, and moves to a larger size class, the project will achieve this objective. For example, Goleta Slough in the Santa Barbara subregion historically supported 724 acres of vegetated marsh and unvegetated flats (size class 500–750 acres (200–300 ha) in Figure 16A) and presently has only 172 acres (70 ha) (size class 100–200 acres in Figure 16A). However, analyses of future projected conditions suggest that there will be 410 acres (160 ha) available at Goleta Slough for wetland migration (size class 400–500 acres in Figure 16A). See Figure 16 for three such examples from each subregion. In addition to the examples in Figure 16, Appendix 4 provides a table of past, present and future wetland acreage for each individual system.

16.A

Santa Barbara Subregion

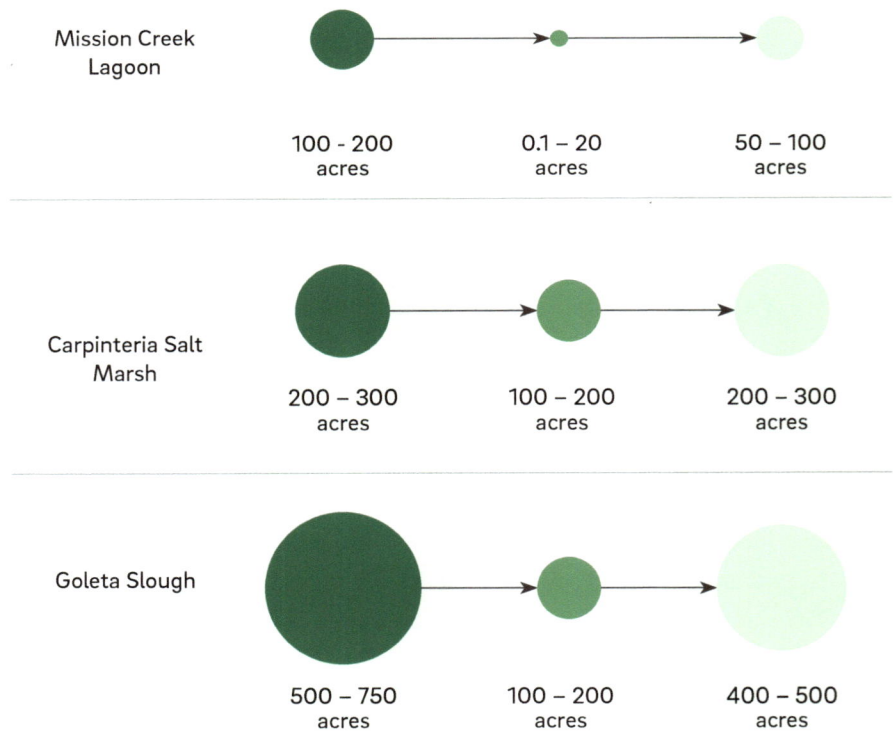

Size of wetland systems over time in the Santa Barbara Subregion.

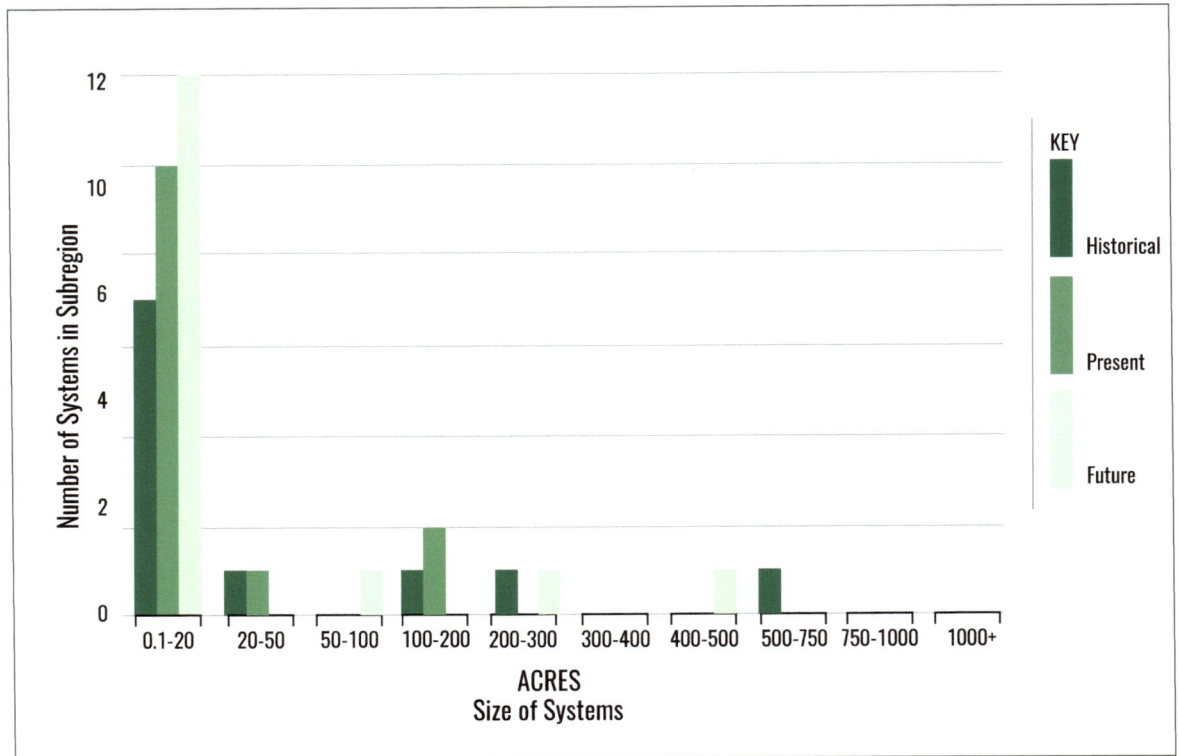

KEY

Historical

Present

Future

Three examples of wetland size over time in the Santa Barbara Subregion. The circles indicate change from historical wetland size class to present day size class, and the potential to change from present day size class to a larger size class in the future.

Subregion	System Name	Historical	Present	Future
SANTA BARBARA	Mission Creek Lagoon	100 - 200 acres	0.1 – 20 acres	50 – 100 acres
	Carpinteria Salt Marsh	200 – 300 acres	100 – 200 acres	200 – 300 acres
	Goleta Slough	500 – 750 acres	100 – 200 acres	400 – 500 acres

16.B

Ventura
Subregion

Size of wetland systems over time in the Ventura Subregion.

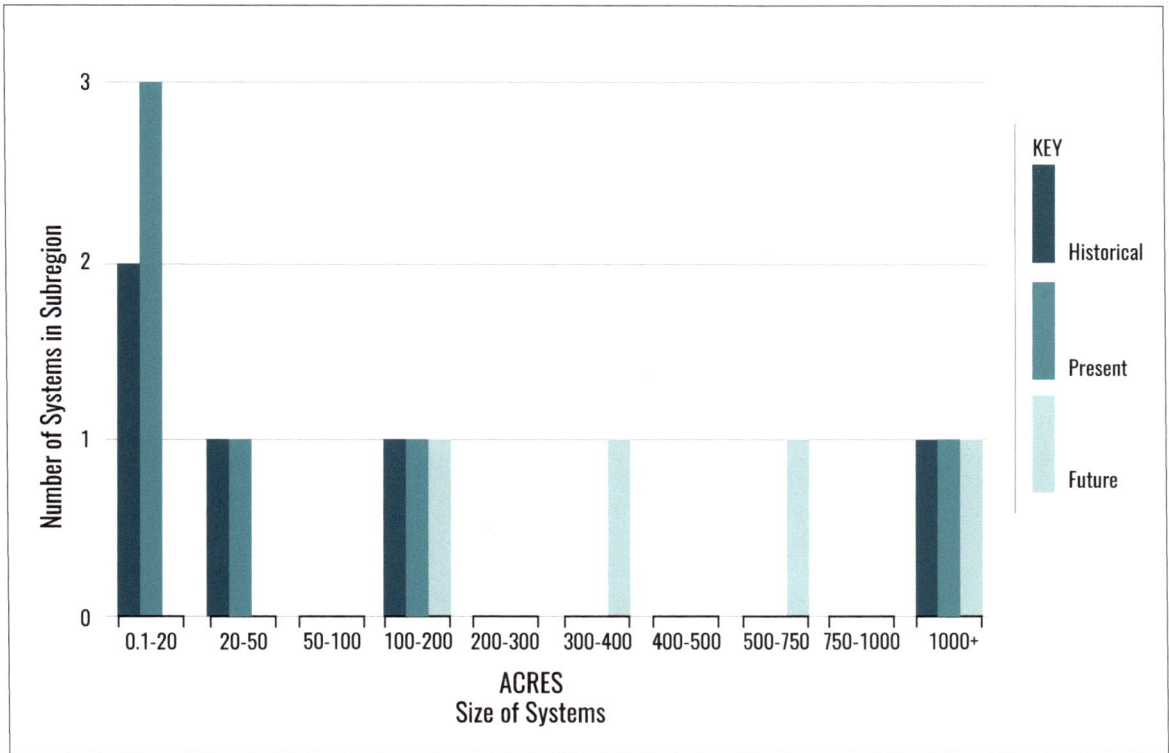

Three examples of wetland size over time in the Ventura Subregion. The circles indicate change from historical wetland size class to present day size class, and the potential to change from present day size class to a larger size class in the future.

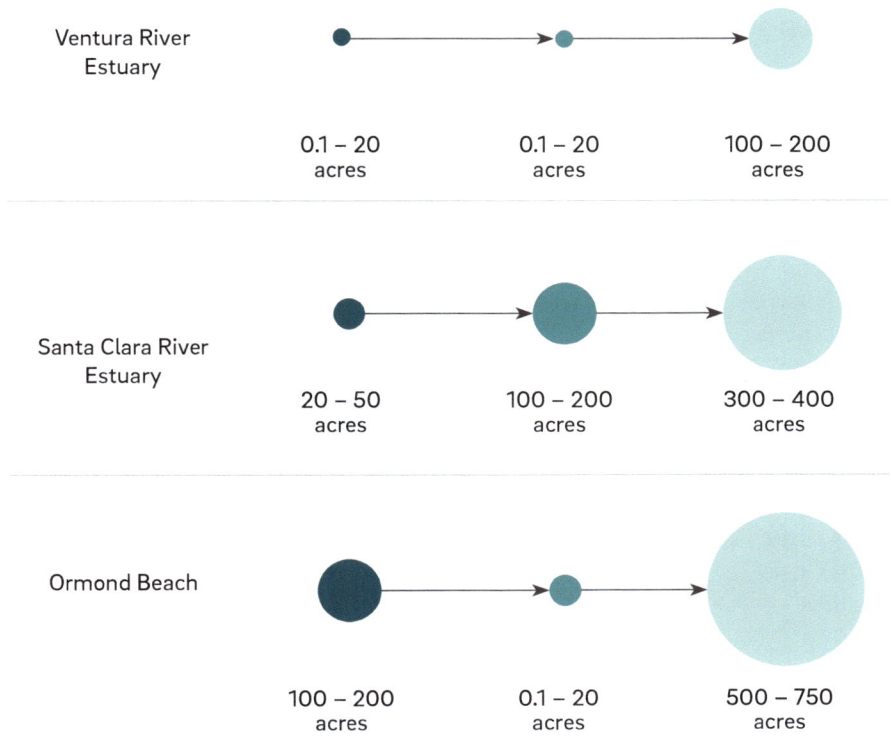

Subregion	System Name	Historical	Present	Future
VENTURA	Ventura River Estuary	0.1 – 20 acres	0.1 – 20 acres	100 – 200 acres
	Santa Clara River Estuary	20 – 50 acres	100 – 200 acres	300 – 400 acres
	Ormond Beach	100 – 200 acres	0.1 – 20 acres	500 – 750 acres

Santa
Monica
Subregion

Size of wetland systems over time in the Santa Monica Subregion.

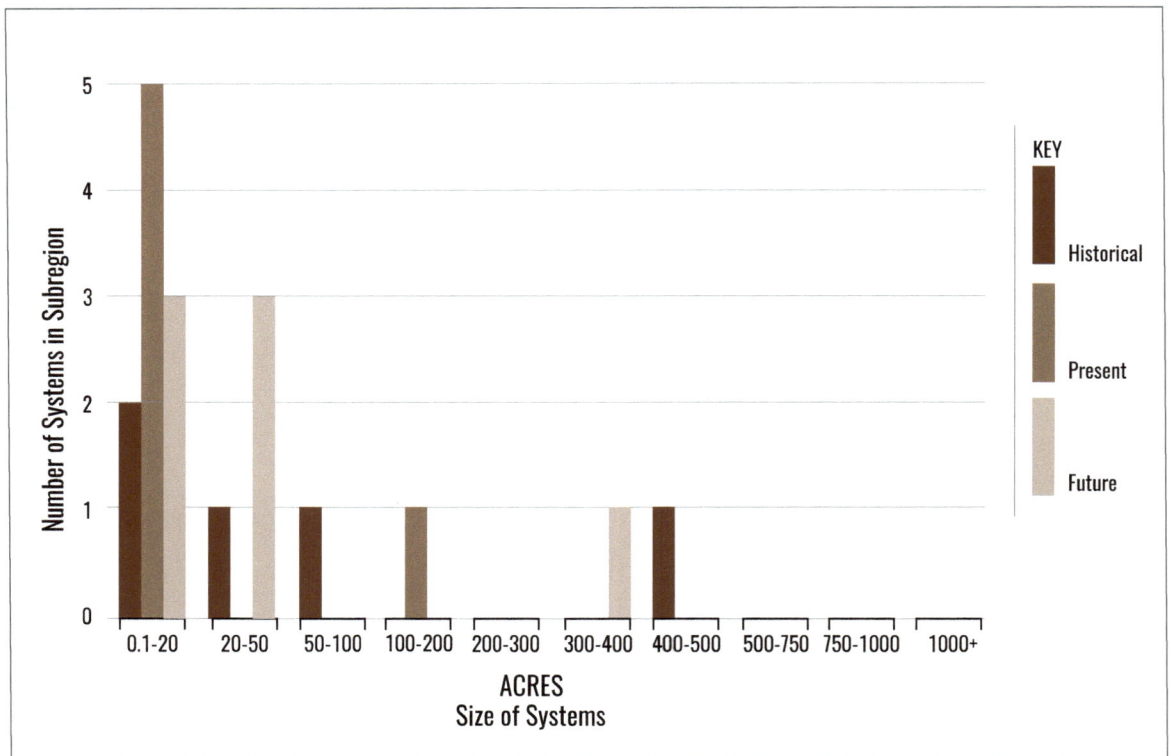

Three examples of wetland size over time in the Santa Monica Subregion. The circles indicate change from historical wetland size class to present day size class, and the potential to change from present day size class to a larger size class in the future.

Subregion	System Name	Historical	Present	Future
SANTA MONICA	Ballona Creek	50 – 100 acres	0.1 – 20 acres	20 – 50 acres
	Ballona Wetlands	400 – 500 acres	100 – 200 acres	300 – 400 acres
	Malibu Lagoon	20 – 50 acres	0.1 – 20 acres	20 – 50 acres

San Pedro Subregion

Size of wetland systems over time in the San Pedro Subregion.

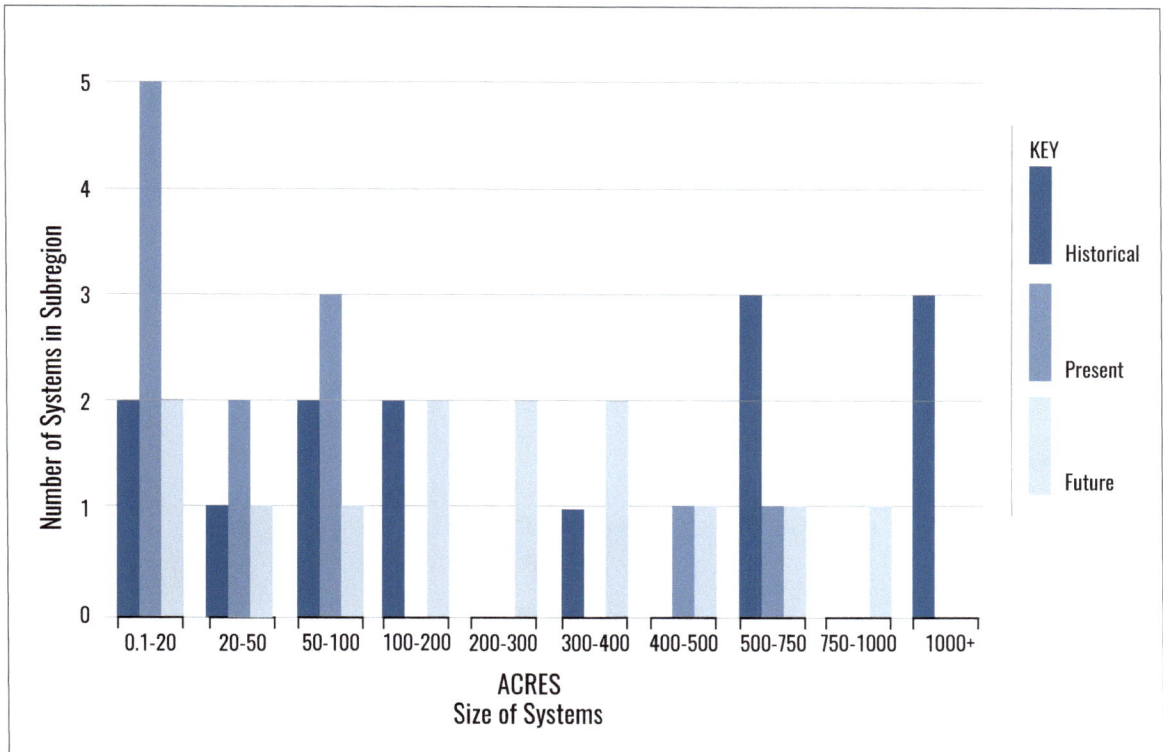

Three examples of wetland size over time in the San Pedro Subregion. The circles indicate change from historical wetland size class to present day size class, and the potential to change from present day size class to a larger size class in the future.

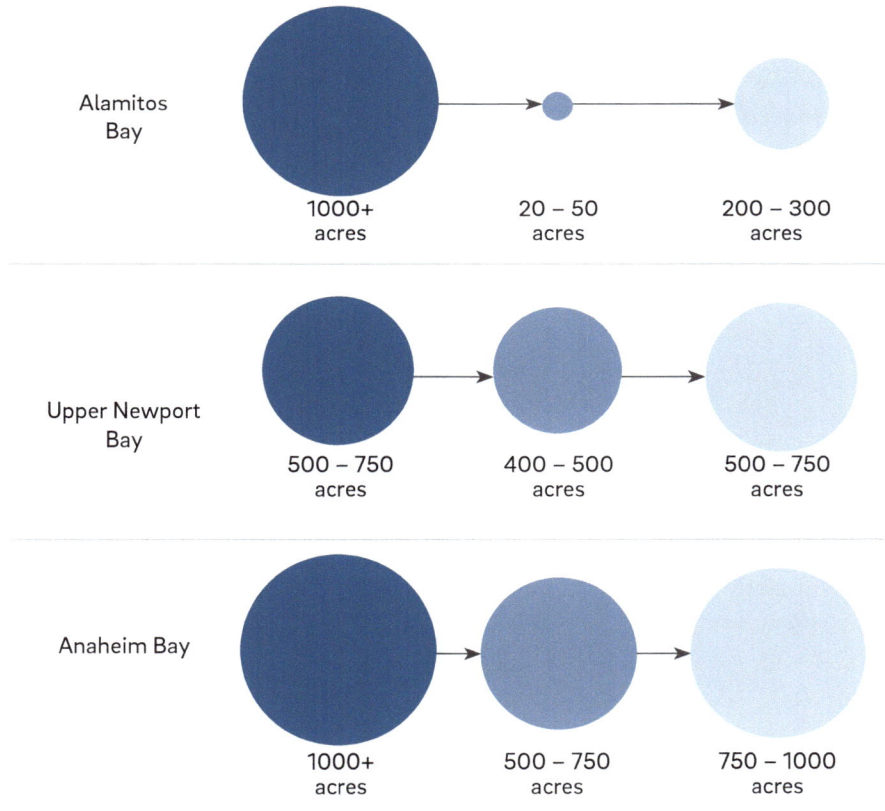

Subregion	System Name	Historical	Present	Future
SAN PEDRO	Alamitos Bay	1000+ acres	20 – 50 acres	200 – 300 acres
	Upper Newport Bay	500 – 750 acres	400 – 500 acres	500 – 750 acres
	Anaheim Bay	1000+ acres	500 – 750 acres	750 – 1000 acres

San
Diego
Subregion

**Size of
wetland
systems over
time in the
San Diego
Subregion.**

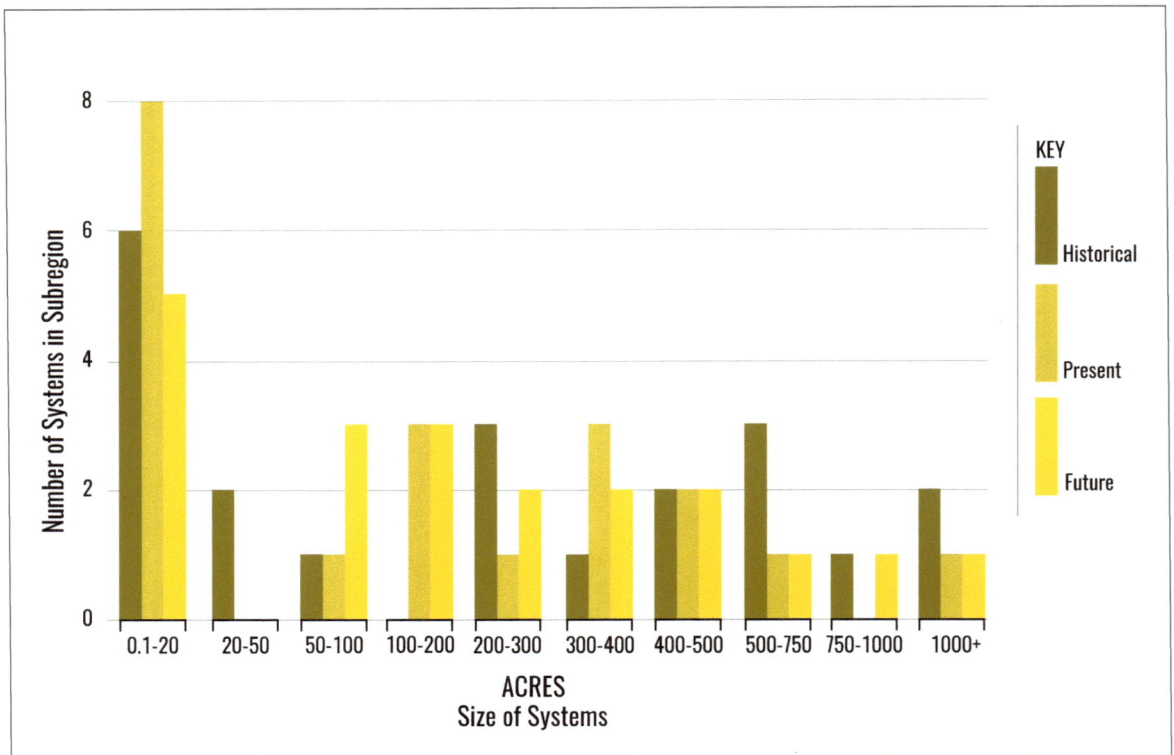

KEY

Historical

Present

Future

**Three
examples of
wetland size
over time in
the San Diego
Subregion.** The
circles indicate
change from
historical
wetland
size class to
present day
size class, and
the potential
to change
from present
day size class
to a larger size
class in the
future.

Subregion	System Name	Historical	Present	Future
SAN DIEGO	Alamitos Bay	300 – 400 acres	100 – 200 acres	500 – 750 acres
	San Diego River Estuary	400 – 500 acres	100 – 200 acres	200 – 300 acres
	Tijuana River Estuary	750 – 1000 acres	500 – 750 acres	750 – 1000 acres

SAN DIEGO RIVER • COPYRIGHT (C) 2002-2018 KENNETH AND GABRIELLE ADELMAN, CALIFORNIA COASTAL RECORDS PROJECT, WWW.CALIFORNIACOASTLINE.ORG

A. Preserve or restore, as appropriate, the historical distribution of wetland archetypes in each subregion as presented in Table 3, below.

B. Increase and maintain connectivity between historically connected wetland fragments, examples of which are presented in Table 3.

A. Rationale for preserving and restoring the historical distribution of wetland archetypes in each subregion:

Wetland archetypes are groups of wetlands that are similar in form, function, and processes. Archetypes provide a general model that can be used to explain how a group of wetlands function and how those wetlands may respond to external pressures or drivers. The rationale of Objective 3A is to realize a distribution of archetypes that is appropriate to their present and future environmental setting guided by their historical form so that they will be more sustainable. Managing systems in a manner appropriate to their environmental setting and allowing them to function more naturally will make them more resilient to stressors, as native species are adapted to the natural environmental conditions in which they evolved, and thus are more likely to persist in those conditions. Maintaining appropriate numbers of similar systems in each archetype will increase the functional redundancy in the region, which promotes wetland diversity and moderates the effects of loss.

Historically, six coastal archetypes can be identified in the region: small creeks, open bays, small lagoons, intermediate estuaries, large lagoons, and large river valley estuaries. The location and historical archetype of each coastal wetland was identified from aerial photographs and historical mapping. Then, each of the 105 wetlands in the region was assigned a current wetland archetype based on contemporary wetland maps.

Many wetlands have been converted from their historical archetype to a new archetype (Table 3 and Figure 17). Archetype conversion indicates a substantial shift in the landscape form and hydrology, and leads to the loss of natural functions that the wetland historically provided. While implementation of all of the other Objectives in the *Regional Strategy 2018* would contribute to historical archetype restoration, Objective 3A specifically identifies coastal wetland systems that have been dramatically altered in form and function. Restoring these coastal wetlands would support appropriate wetland functions for native species in each environmental setting.

There are coastal wetlands in Southern California that have maintained their historical archetype classification even though they are actively managed to function as a different archetype. For instance, Intermediate Estuaries whose tidal inlets are managed to remain open and have not necessarily been altered in form or fragmented, have not undergone an archetype shift. However, the need to restore appropriate connectivity to the ocean and watershed, in situations such as these, is the subject of Objective 6: Hydrological Connectivity.

Table 3. Historical and present wetland archetypes including current wetland fragments. Wetlands that are presently fragmented may have opportunities to reconnect fragments and improve wetland function. Some of these wetland fragments are separated by federal flood channels and would be difficult to restore full hydrologic connection, but the possibility of restoration should be explored when possible.

Subregion	Historical Archetype (the Objective)	Present Archetype	Wetland	Fragments
Santa Barbara	Small Creek	Small Lagoon	Tajiguas Creek	N/A
	Intermediate Estuary	Small Creek	Mission Creek Lagoon	N/A
	Small Creek	Small Lagoon	Andree Clark Bird Refuge	N/A
	Small Creek	Small lagoon	Sanjon Barranca	N/A
Ventura	Large River Valley Estuary	Open Bay/Harbor & Fragmented River Valley Estuary	Ventura River Estuary, Santa Clara River	Ventura Marina, McGrath Lake, & Santa Clara River
	Small Creek	Small Lagoon	San Buenaventura	N/A
	Large River Valley Estuary*	Fragmented River Valley Estuary*	Ormond Beach*	
Santa Monica	Intermediate Estuary	Open Bay/Harbor & Fragmented River Valley Estuary	Ballona wetlands	Marina Del Rey, Ballona Lagoon, Ballona Creek, Ballona Wetlands, & Del Rey Lagoon
San Pedro	Large River Valley Estuary	Fragmented River Valley Estuary	Los Angeles River	Los Angeles River Channel
	Large River Valley Estuary	Fragmented River Valley Estuary	Los Cerritos	Alamitos Bay, Los Cerritos Wetlands, Los Cerritos Channel, & San Gabriel River Estuary
	Large Lagoon	Large Lagoon & Open Bay/Harbor	Anaheim Bay (Seal Beach wetlands)	Anaheim Bay, Huntington Harbor, & Bolsa Chica Channel
	Large Lagoon	Small Creek & Large Lagoon	Bolsa Chica	Bolsa Chica Lagoon, Bolsa Bay, & Wintersburg Channel
	Large River Valley Estuary	Fragmented River Valley Estuary	Santa Ana River Mouth	Huntington Beach Wetlands, Santa Ana River, & Santa Ana River Wetlands
San Diego	Large River Valley Estuary	Open Bay/Harbor	Oceanside Harbor	N/A
	Large Lagoon	Small Creek	Loma Alta Slough	N/A
	Large River Valley Estuary	Open Bay/Harbor, Intermediate Estuary, & Large River Valley Estuary	Mission Bay	Mission Bay, North Mission Bay Wetlands, & San Diego River
	Open Bay/Harbor	Open Bay/Harbor & intermediate Estuary	San Diego Bay	San Diego Bay, Sweetwater Marsh, & Otay River Estuary

* Archetype Objective for Ormond Beach is not the same as all of the other Objectives (see Rationale A below)

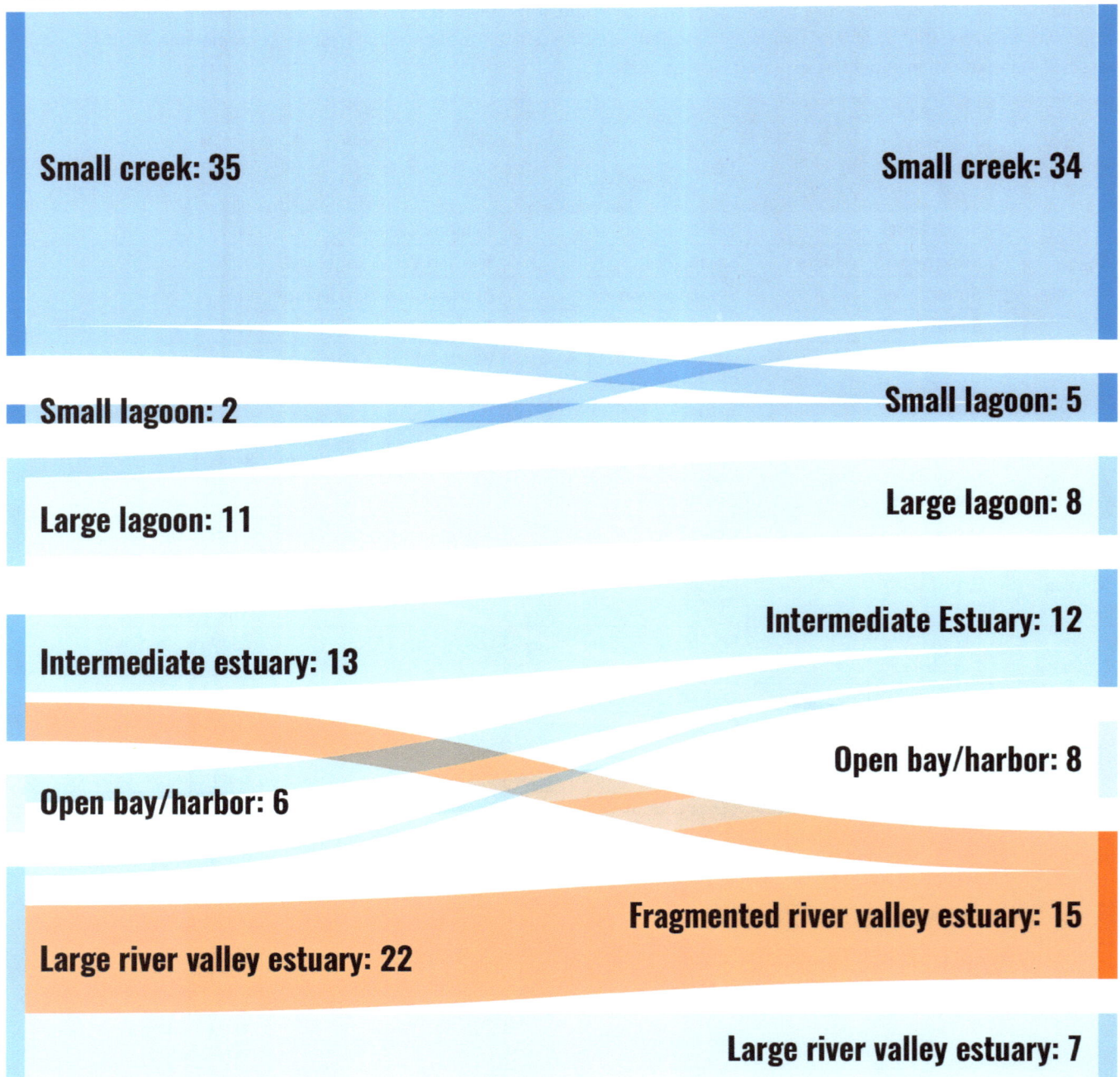

Figure 17. Archetype change across coastal wetlands in Southern California with historical archetype on the left and present archetype on the right. Historical archetype could not be determined for certain systems, and these systems were excluded from this figure.

EXCEPTIONS TO THE RULE

As with all classification systems there are outliers that provide exceptions; Ormond Beach and Mugu Lagoon are two of them. Ormond Beach was historically part of the Large River Valley Estuary of the Santa Clara River Delta whose distributary channels migrated across the large, flat Oxnard Plain leaving traces of wetland features as the river created, and then abandoned, a series of tidal inlet locations along the coast (Stillwater Sciences 2011, Beller et al. 2011). As recently as perhaps 200 to 500 years ago, the Santa Clara River may have shifted its tidal inlet from near Point Hueneme to its present location. The hydrologic connection between wetlands, creeks and the ocean were intermittent—Calleguas Creek was hydrologically connected to Mugu Lagoon through shallow sloughs and sheet flow only during floods. Today, the shoreline of the Oxnard Plain is composed of the freshwater-brackish, intermittently closed estuaries of the Ventura and Santa Clara rivers; the non-tidal lagoon complexes marking former

Santa Clara River mouths such as Ormond Beach, and the large, more tidally-influenced wetland system at Mugu. Ormond Beach is therefore a part of a Fragmented River Valley Estuary, with several parcels of historical wetland in close proximity but hydrologically disconnected from one another. Other parts of the Fragmented River Valley Estuary, such as the present Santa Clara River, are more distant from Ormond Beach and separated by the development of the city of Oxnard. They are wetland systems in their own right now. Due to the infeasibility of reconnecting these other fragments to Ormond Beach now, Ormond Beach is the only coastal wetland in Southern California where the WRP recommends future restoration to result in the creation of an archetype different from its historical classification. Given its present environmental setting, the WRP proposes the most appropriate archetype for a restored Ormond Beach is an Intermediate Estuary, assuming that sufficient freshwater flows are available.

MUGU LAGOON • COURTESY OF GOOGLE EARTH

B. Rationale for increasing and maintaining connectivity between historically connected wetland fragments:
Where the wetland has been relatively unaltered, the Objective is to maintain the current archetype and maintain or increase the connectivity within the wetlands. Connectivity provides the connections, space, and physical and biological gradients needed for species to move in response to changing conditions. Connectivity allows organisms to escape unfavorable conditions, take advantage of redistributed or newly available resources, recolonize areas after a disturbance, and exchange genes between populations. As a result, habitats can shift, species can adapt, and communities can reorganize as conditions change.

In other cases, wetlands have been fundamentally altered. For example, many of the Large River Valley Estuaries have been fragmented by diking, draining, and development, which has left the remaining wetlands isolated and disconnected from each other. Reconnection may range from managing fragments in complementary ways, planning restoration more comprehensively for all fragments (even when some fragments are not in public ownership or hydrologically connected), creating physical corridors for species migration, creating hydrological connections through pipes and culverts, to fully restoring open channels and breaching levees. Reconnecting currently fragmented pieces will improve their present functioning and resilience, and restore conditions under which the native species evolved, to which they are adapted, and under which they are likely to persist. In cases where the system is too disconnected, as in Ormond Beach described on the previous page, a more appropriate archetype is proposed for the present day environmental setting.

The Santa Ana River system is an example of one of these fragmented wetlands that has been fundamentally altered (Figure 18). Historically the Santa Ana has switched its course many times. For instance, in the 1820s, large floods caused it to flow to the ocean at Newport Bay for a period of time, creating Balboa Island in the process (City of Santa Ana 2006). From the 1920s to 1990s many projects were completed to try to control the river, from construction of the Bitter Point Dam to bypass Newport Bay and flow directly to the ocean, to numerous levees and dams. These efforts culminated in the Santa Ana River Mainstem Project of the 1980s, which created a straightened, concrete trapezoidal channel for 30 miles of the lower reaches of the river, and the Seven Oaks Dam of the 1990s which significantly reduced the peak flood flows. As a result of these flood control projects, the river today functions mainly as a stormwater channel, cut off from its floodplain and adjacent wetlands. While it is unrealistic that the fragments shown in Figure 18 will be hydrologically reconnected to the fullest extent (i.e., fully tidal with no concrete channel), there are many opportunities in this system to reconnect the fragmented wetlands at other levels. For instance, wetland managers could consider and include all of the wetland fragments when planning future restoration designs to not preclude reconnection possibilities in the distant future. Additionally, management efforts in adjacent, hydrologically disconnected wetland fragments could be complementary by coordinating habitat and species goals, and even creating physical corridors for species migration between the fragments. Finally, hydrological connection could be made through pipes and culverts—even if only muted tidal exchange is achieved, the habitat and species benefits would be significantly improved compared to the present day fragmentation.

Historical Wetlands

Contemporary Wetlands

Huntington
Beach
Wetlands

Santa Ana River

Santa Ana
Wetlands

Figure 18. Historical system with opportunities for reconnecting fragments at the Santa Ana River mouth. In this system, the contemporary fragments include Huntington Beach Wetlands (brown outline), the Santa Ana River (light brown outline) and the Santa Ana Wetlands (yellow outline).

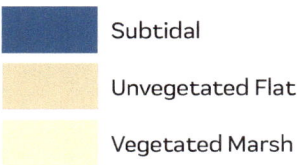

N

1/4 mile

Subtidal

Unvegetated Flat

Vegetated Marsh

Management Strategies

To achieve Objective 3, the WRP recommends three Management Strategies focusing on modifying the landscape, both within the watershed and at the ocean, to reconnect water and sediment to coastal wetlands and to restore cycles of tidal inundation (Figure 19).

First, the construction of dams and berms and the channelization of creeks has disrupted the natural hydrologic and sediment pathways in many wetlands in Southern California. To allow coastal wetlands to function and evolve requires the reconnection of water and sediment pathways to the wetlands in terms of volume, timing, and location. Such efforts in the coastal wetlands should be related to Management Strategies higher up in the watersheds where sediment and water are detained by dams, detention basins, and other such structures.

Second, the natural tidal inlets of Southern California coastal wetlands include those that are permanently open and those with varying degrees of intermittent opening. In wetlands, the presence of an intermittently opening tidal inlet has been an important driver for the ecological evolution of habitats. Since the late 1800s, the functioning of many inlets has been altered by managed breaching, altered sediment loading and tidal prism, and the construction of structures such as jetties. The movement of inlet channels has also been impeded by bridges, roads and berms crossing the wetlands. Along with a reduction of stressors from the watershed, the restoration of more natural inlet opening conditions would restore many of the natural functions of intermittently-opening estuaries that have been lost. For a full discussion on intermittently-opening wetlands, see pages 54–55 . Restoring more natural opening conditions requires changes in the management and structure of the tidal inlet. However, difficult tradeoffs must be evaluated for each project to determine the gain in ecological function with inlet restoration versus the potential loss in functions that could occur as a result of poor water quality or flooding. Coastal resource managers are trying to find the right approach for making these decisions (Appendix 5 and Appendix 6). Goal 4 (page 104) provides a thorough discussion on the scientific research needs for intermittently-opening estuaries.

> **Management Strategy 5: Remove barriers to reconnect wetlands.** The goal of this Management Strategy is to facilitate species movement, natural sediment transport and hydrological flows between wetland fragments and from river channels into wetlands. Species movement between wetlands is often interrupted by the presence of roads, berms and development. Creating wildlife corridors and safe methods for crossing obstacles, such as wildlife bridges, will facilitate this movement. Built infrastructure in the wetlands also impedes natural water flow. The best way to restore the natural functions that create and maintain wetlands is to remove the physical impediments to water and sediment flow that have been built. In situations where flood risk management infrastructure inhibits water flows, culverts or gates could be installed to allow water to flow from creeks into adjacent wetlands. Enhancing sediment movement may require more than culverts or tide gates, since significant sediment loads can move with extreme flood events. In these cases, breaching or lowering berms would facilitate sediment transport. Additionally, new berms could be constructed farther back from the wetlands in order to maintain the same level of protection to the urban and residential areas they were originally intended to serve.

LOS CERRITOS WETLAND • PHOTO BY AKF2006, COURTESY OF CREATIVE COMMONS

Management Strategy 6: Allow tidal inlets to open and close naturally. In the case of tidal inlets that have natural beach berms that form and break throughout the year based on wave action and freshwater flow, inlet breaching is a commonly-practiced management strategy. However, inlet breaching might not be the most effective management action for wetland restoration and management. In some cases, allowing the wetland to naturally fluctuate between open and closed, and to exhibit natural fluctuations in flooding, dissolved oxygen, and stratification, will restore functions such as species support that are important to land managers. An inlet management plan that quantifies the ecological tradeoffs for breaching or not breaching the inlets of closed systems should be developed for each site with a managed inlet. The inlet management plan should weigh ecological drivers, flooding, and water quality issues.

Management Strategy 7: Modify or remove structures to restore inundation regime. Jetties and other structures have been constructed in many places to hold tidal inlets open, transforming the inundation regime to be continuously open to the ocean throughout the year. In addition, the movement of channels inside the estuary has been restricted by the presence of bridge crossings and other such structures. Where opportunities exist, structures should be modified or removed to facilitate a more dynamic tidal inlet and channel network that can move in response to changing conditions, opening and closing, and moving with a more natural frequency. In case where shoreline protection is needed, consider employing a "living shoreline" approach that is appropriate for the needs and conditions of the site (page 69).

Objective Tracking

If a project can restore a system closer to its historical archetype, the system's structure and processes should match the historical archetype criteria. The archetype assignments for example system are found in Table 3 and the processes and structure of the historical archetypes are presented in the archetype conceptual models (pages 18–23). Each project will report the current archetype of the coastal wetland in which the project is located, as well as the projected archetype if the project will result in a change.

Project monitoring should assess the degree of reconnection between system fragments. For instance, the presence of wildlife corridors or the degree of hydrologic connectivity between fragments could be assessed by site inspection and/or Geographic Information Systems (GIS) mapping technology. If a large wetland has been fragmented into several smaller wetlands, the ability of species to move, or the flow of water between those systems will be reduced or non-existent.

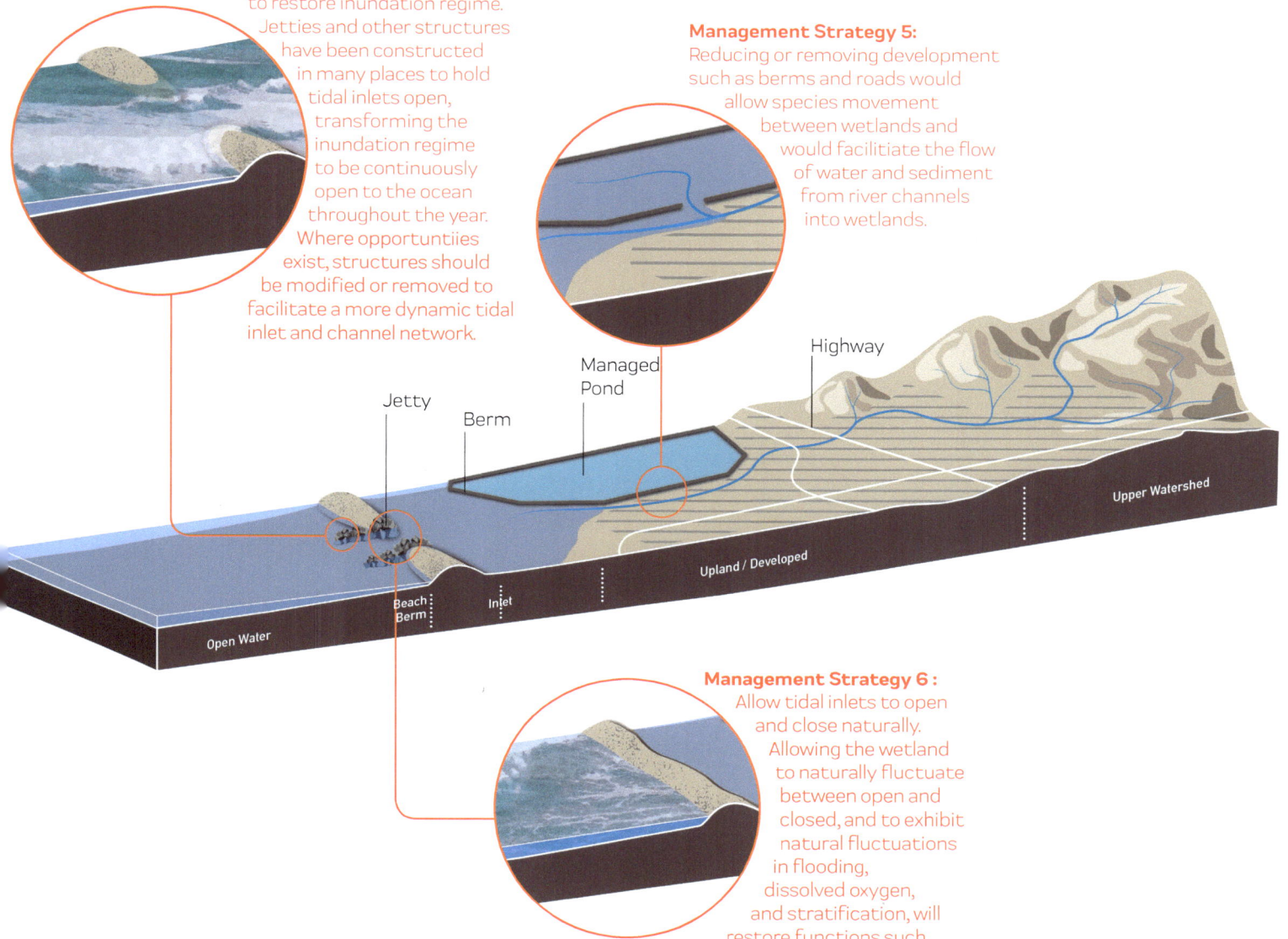

Management Strategy 7:
Modify or remove structures to restore inundation regime. Jetties and other structures have been constructed in many places to hold tidal inlets open, transforming the inundation regime to be continuously open to the ocean throughout the year. Where opportuntiies exist, structures should be modified or removed to facilitate a more dynamic tidal inlet and channel network.

Management Strategy 5:
Reducing or removing development such as berms and roads would allow species movement between wetlands and would facilitiate the flow of water and sediment from river channels into wetlands.

Management Strategy 6 :
Allow tidal inlets to open and close naturally. Allowing the wetland to naturally fluctuate between open and closed, and to exhibit natural fluctuations in flooding, dissolved oxygen, and stratification, will restore functions such as species support that are important to land managers.

Highway

Managed Pond

Jetty

Berm

Upper Watershed

Upland / Developed

Beach Berm

Inlet

Open Water

Figure 19. Conceptual diagram depicting the Management Strategies 5–7.

INTERMITTENTLY OPENING ESTUARIES

A key controlling factor of the wetlands of Southern California is the nature of their connection to the ocean, controlled by the characteristics of the tidal inlet. Tidal inlets range from permanently open, through varying degrees of intermittent opening, to closed. The frequency, timing, and duration of inlet closure will affect the elevation of water and salinity within the wetlands which, in turn, will have significant impacts on vegetation. The tidal inlet provides passage for sediment, nutrients and fish and each will be affected by the degree, timing and frequency of closure. Over the last century, many of the tidal inlets have been modified by the construction of jetties for both transportation infrastructure and flooding purposes. These modifications tended to stabilize the inlets and have maintained them in a more open condition. Restoring more natural tidal inlet conditions would help restore many of the natural functions of these wetlands and is a focus of the *Regional Strategy 2018*.

Many of Southern California's wetlands have, or historically had, dynamic connections with the ocean that varied seasonally reflecting annual patterns of precipitation and river discharge, as well as multi-year patterns of wet and dry years. The dynamic nature of these inlets is an important characteristic of many Southern California wetlands that was captured in every archetype classification except Open Bays/Harbors. Wetlands that are defined by dynamic tidal inlets are variously referred to as Intermittently Opening Estuaries (IOE) and Bar-Built Estuaries (BBE), and in the archetype classification mostly occur in the "Intermediate Estuary" category. The Intermediate Estuary archetype reflects a balance between the river and wave forces on the continuum of tidal inlet conditions. All of the wetland archetypes lie on a continuum of inlet state, ranging between mainly closed to perched to mainly open, depending upon the balance of river flow to wave energy, from fluvial-dominated river mouth estuaries, to wave-dominated lagoons (Gleason et al. 2011). The tidal

inlet state is also dependent upon wetland size. A Large River Valley Estuary system, where flow persists for weeks or months, is more likely to stay open and less likely to close compared to a Small Creek, which only has flow following rainfall, given the same wave exposure. Similarly, a Large Lagoon system is more likely to stay open and less likely to close compared to a small lagoon in similar circumstances.

Each of the seven wetland archetypes are characterized by tidal inlet conditions that vary along a continuum between mainly closed to the ocean to always open to the ocean. This reflects the variable nature of river flow in relation to wave energy and tidal prism. The balance of forces between river flow, tides and wave energy and their effect on inlet condition is an important factor in determining wetland archetype. The tidal inlet continuum is shown in Figure 20, which illustrates the dominant forcing for the system, whether it is tidal-dominated (open bay/harbor), fluvial-dominated (rivers and creeks) or wave-dominated (lagoons).

The forcing factors determine where on the continuum the inlet state lies, and hence the habitat types, flora, and fauna supported by each estuary (Jacobs et al. 2011). The inlet state will have significant impacts on the water quality of the estuary, dependent upon the tidal state when closing occurred and the amount of freshwater dilution from the river. Salt water may be trapped when the inlet closes, and this may lead to stratification as fresh water continues to flow over the trapped oceanic water. This trapping and stratification may increase temperature and salinity and decrease dissolved oxygen. In these systems, anoxic events may be driven more by inlet closure than by seasonality.

The water level within the closed system is sensitive to the water balance driven by river flow, seepage, overtopping, and evaporation. If the system closes when the water level is relatively high, and there is subsequent river flow, then flooding may be an issue. If the system closes when the water level is relatively low during the summer when evaporation is high, then water quality may deteriorate, creating hypoxic zones and providing breeding habitat for mosquitoes. The change in mean water level will also affect the inundation regime of the wetlands—the depth and duration of flooding. This shift in inundation, together with the salinity variations, will have significant impacts on wetland vegetation.

The characteristics of the inlet , whether it is open, perched or closed, will affect not only the passage of water but also the passage of sediment, nutrients, and biota. Changes in the timing of closure may have significant impacts on particular species as it interacts with events in their life-history. Migration opportunities may be significantly reduced or occur at different times of the year according to the type of water year and the amount of wave energy. For instance, during a drought, the closure period may be delayed by months or even years longer and so the migration period may be shorter or unavailable. Additionally, changes in water chemistry in the estuary may limit mixing zones for salmonids to adjust from salt to fresh water, which may impose

a barrier to passage even when there is not the complete physical closure of the inlet.

There are benefits to inlet closure. The ecosystem within these coastal wetland systems has evolved to accommodate inherent variability and to living under stressful conditions. The accumulation of freshwater lenses may benefit many plant species, even saltwater plants. Stable conditions during closure can promote the growth of submerged plants (Riddin and Adams 2008), which may provide a foundation for epiflora and epifauna (Davies 1982) and refuge habitat for juvenile fish (Whitfield 1984). There are benefits to physical processes—fine sediment that enters the estuary during relatively small floods when it is closed is more likely to accrete rather than be flushed out to the ocean. This sediment accretion process is vital for wetlands to keep pace with sea-level rise. Higher water levels in the estuary can increase the extent of wetland habitat (Perissinotto et al. 2010). Groundwater elevations adjacent to the estuary may be elevated to a greater degree by the longer duration of high water levels in the estuary.

Figure 20. Relative influence of river, tide, and wave energy on tidal wetlands (Adapted from Gleason et al. 2011).

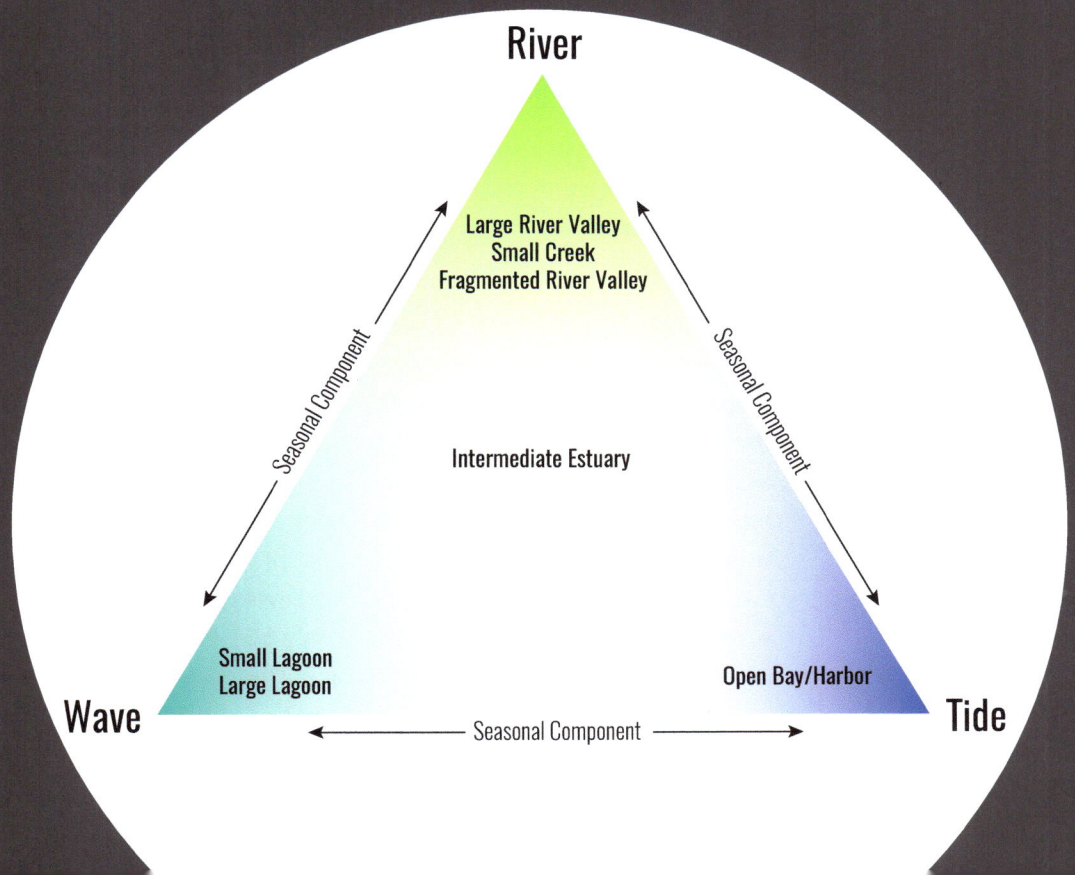

Restore or maintain the coastal wetland composition, represented by the historical archetype habitat profiles, in at least 50% of the systems within a given archetype across a subregion, as represented in Table 4.

Table 4. Recommended intertidal habitat profiles by subregion and archetype. These are the habitat profiles that should be realized with 24 inches of sea-level rise.

Subregion	Archetype	Unvegetated Flat	Vegetated Marsh
Santa Barbara Coast	Intermediate Estuary	22%	78%
	Large River Valley Estuary	48%	52%
Ventura Coast	Intermediate Estuary	27%	73%
	Large River Valley Estuary	33%	67%
Santa Monica Bay	Intermediate Estuary	27%	73%
San Pedro Bay	Large Lagoon	24%	76%
	Large River Valley Estuary	18%	82%
San Diego Coast	Intermediate Estuary	14%	86%
	Large Lagoon	72%	28%
	Large River Valley Estuary	35%	65%

Note: Proportions are shown only for unvegetated flat and vegetated marsh. The proportion of subtidal channel should be sufficient to support the flats and marsh.

Rationale: This Objective is intended to approximate the historical diversity (the variety of landscape features) and complexity (the spatial configuration and interactions between the features) of habitat types across wetlands in the region. Diversity and complexity help maintain the variability necessary for species adaptation and evolution by supporting a range of responses to a heterogeneous and dynamic environment. Diverse habitat mosaics can support more niches, bolstering biodiversity and promoting alternative life-history strategies. Complex landscapes, supporting a variety of microhabitats, provide individuals with opportunities for acclimation or refuge during disturbance events and extreme conditions.

The proposed proportion of unvegetated flats and vegetated marsh was calculated from an analysis of the historical mapping for the region (Figure 21 shows an example of historical mapping at Batiquitos Lagoon). Habitat profiles were averaged by subregion and archetype for the archetypes historically present in each subregion. Small creeks and small lagoons, as well as intermediate estuaries in some subregions, were not analyzed as the snapshot provided by historical mapping shows them to be highly variable in their habitat diversity. Subtidal habitat was not included in the proposed habitat proportions because with sea-level rise this type of habitat would increase naturally, and to reverse this trend would require filling or diking of open water. The WRP set the Objective for only 50% of systems to keep the Objective achievable, yet ambitious, based on the best-professional judgement of the Science Advisory Panel.

Figure 21. These maps shows an example of the historical (left) and contemporary (right) distribution of wetland habitats at Batiquitos Lagoon.

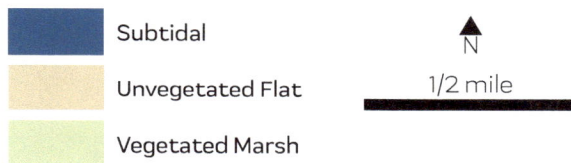

- Subtidal
- Unvegetated Flat
- Vegetated Marsh

N
1/2 mile

While the proposed habitat distributions for this Objective reflect the historical conditions, these distributions should only act as a guide for wetland habitat design in the future. When designing a project to be resilient to sea-level rise, the design should model the proposed habitat distributions with 24 inches of sea-level rise. The result of this approach will be a habitat design that is built with slightly higher elevations to accommodate the change in water elevations. Since wetland restoration projects take many years to design and build, projects that are designed today should be built to result in these habitat distributions with sea-level rise.

Management Strategies: The fragmentation of coastal wetlands through the construction of levees and the alterations of tidal inlets has fundamentally changed the distribution of habitats within the wetlands. Some habitats, like salt flats and alkaline marshes have been disregarded in many wetland restoration activities. For some salt flats, tidal inlet management has prevented the natural wetting and drying of the flats ultimately converting those flats into salt marsh habitats or disturbed upland. For some alkaline marshes, an increase in freshwater and sediment inputs from the watershed have converted those alkaline vegetation species to riparian, freshwater species. Similarly in shallow subtidal areas, activities such as dredging and urbanization have impacted invertebrates and other species living on the soft bottom substrate and/or reef/hard-bottoms.

Management Strategies to restore habitat diversity should focus on the restoration of wetland form and function in order to restore an appropriate habitat mix, and should focus on particularly unique habitats.

Management Strategies 5, 6, and 7 (see Objective 3), and the following:

NAVANAX IN NEWLY PLANTED EELGRASS IN UPPER NEWPORT BAY •
PHOTO COURTESY OF ORANGE COUNTY COASTKEEPER

Management Strategy 8: Protect existing natural salt flats and their supporting hydrological regime, while also protecting anthropogenic salt flats where it can be demonstrated they have value that other habitats within the system cannot support. Protecting existing salt flat habitats involves preventing future disturbance (e.g. not allowing heavy machinery being driven or future construction on the flats) and/or restoring the natural tidal inlet dynamics to allow for seasonal wetting and drying of the flats. The WRP has analyzed the historical distribution, outlined potential functions of salt flat habitats, and developed a salt flat typology (Appendix 7). The WRP should pursue the development of quantitative restoration Objectives for these habitat types in a future phase of the Rregional Strategy (Goal 4).

Management Strategy 9: Protect existing shallow subtidal habitats associated with coastal wetlands. Protecting existing subtidal habitats involves preventing future disturbance (e.g. dredging or future construction on the subtidal habitat). The WRP has identified key subtidal habitats that should be protected in the meantime (Appendix 8). The WRP should pursue the development of quantitative restoration Objectives for these habitat types in a future phase of the Regional Strategy (Goal 4).

SUBTIDAL HABITATS

Our understanding of the historical and present distribution of subtidal habitats is limited by the fact that these habitats have been typically mapped as uniform open water, with little detail on key sub-habitat types. In reality, subtidal habitats can range from sessile species attached to harbor pilings to extensive eelgrass beds in estuaries. While subtidal habitat types occur across the whole Southern California Bight, they are diverse and dynamic. Their key characteristics, persistence over time, and biological composition can vary among wetlands, seasons and from year-to-year as water temperature, turbidity, and salinity vary. This diversity is in part a reflection of the diversity of wetland types found in Southern California.

The subtidal habitats are critical in terms of food web, fisheries, and numerous other ecosystem services and functions. Ecologically and commercially important fish rear in warm estuarine waters, rich with small invertebrates.

Many of these species reproduce in the ocean, then as small juveniles migrate into estuaries to rear, and as larger juveniles migrate back to the ocean. Subtidal habitats also serve as migration corridors. Submerged aquatic vegetation is a critical habitat as it serves as nursery habitat, foraging grounds, and refuge for fish; as well as removing nutrients and suspended sediments from the water column, and improving water clarity.

There has been drastic habitat type conversion and degradation in subtidal habitat primarily due to dredging activities and urbanization. In many places the subtidal zone has been deepened to allow navigation. Sedimentation regimes in coastal embayments have been highly altered as a result of upstream development and construction, which impacts subtidal habitats.

See Appendix 8 for more information on subtidal habitats.

SALT FLATS

Salt flats—also known as salt pans (or pannes), *salinas,* alkali flats, playas, and sabkhas—are unvegetated seasonal wetlands characteristic of arid, semi-arid, and Mediterranean climate coastal environments such as Southern California. They form in areas disconnected from regular tidal inundation and where water ponds—perhaps above normal tides (supratidal) or behind an ocean barrier such as sand dunes or a closed estuary mouth (intertidal). High salt concentration in the soils, a result of high evaporation rates, keeps them unvegetated. They experience large fluctuations in salinity and inundation—parts of the year they are dry flats and other times they are shallow ponded areas. As a consequence their ecosystem functions and services vary during the year and between years.

When flooded, salt flats can support foraging for resident and migratory birds: dabbling ducks and shorebirds can feed on invertebrates, invertebrate larvae, and the occasional small fish, while diving birds such as grebes, cormorants, and ruddy ducks can feed in deeper water. Drying salt flats can provide breeding habitat for the state and federally endangered California least tern and federally threatened western snowy plover, in addition to resident birds such as black-necked stilts and American avocets. When dry, salt flats can support roosting and refuge for birds able to congregate safely in the large open space, as well as corridors for traveling mammals and habitat for invertebrates such as tiger and rove beetles and micro-crustacean and aquatic insects such as water boatman and brine flies.

Salt flats were historically present in approximately one-quarter of Southern California estuarine systems. They were found across the Bight, from Goleta Slough to the Tijuana River Estuary, and covered more than 3,000 acres (1,210 ha) or about 10% of the total estuarine area. The largest salt flats (between 150 and 1,000 acres [60–400 ha]) were found in Batiquitos, San Elijo, Buena Vista, and Agua Hedionda lagoons, at Goleta, and Mugu Lagoon. Salt flats have experienced dramatic changes since the 19th century, with losses of approximately 80% of total area. However, there are still a few salt flats in the California Bight region, for example in Devereux Slough, San Elijo Lagoon, and Tijuana River estuaries.

The lack of knowledge of the formative processes, historical and contemporary distribution, and ecosystem functions and services of salt flats in the region means we are currently unable to identify appropriate regional Objectives for salt flat management. They are an important component of overall estuarine and transition zone habitat diversity in Southern California and appear to be under-represented in current systems.

See Appendix 7 for more information on salt flats.

Objective Tracking:

The habitat types used in the *Regional Strategy 2018* Objectives are: subtidal, unvegetated flat, vegetated marsh, and wetland-upland transition zone (pages 64–65 and Table 5). Measure the area of each of the coastal wetland habitats. Each project will provide baseline data so that pre- and post-project habitat distributions can be compared.

Table 5. Crosswalk between Cowardin et al. 1979 wetland habitat types and the wetland habitat types used in the *Regional Strategy 2018*. These are the habitat profiles that should be realized with 24 inches of sea-level rise.

Regional Strategy Habitat Types	Cowardin et al. 1979 Wetland Habitats			
Subtidal	Unconsolidated bottom	Rock bottom	Aquatic bed	Reef
Unvegetated flat	Unconsolidated shore (including salt flats)			
Vegetated marsh	Emergent	Scrub-shrub	Forested	
Transition zone	N/A (see Objective 5 and Appendix 9 for more information)			

SALTMARSH (VEGETATED FLAT), MUDFLAT (UNVEGETATED FLAT), TIDAL CHANNELS AND OPEN WATER (SUBTIDAL) AT KENDALL-FROST MISSION BAY MARSH RESERVE
PHOTO BY LOBSANG WANGDU, COURTESY OF UC NATURAL RESERVE SYSTEM

A. Protect all existing natural areas of wetland-upland transition zones (Figure 22) from the wetland boundary/edge (depending on topography) out to 1,600 feet (500 m). New structures within transition zones should be minimal, not impede wetland migration, and potentially removable.

B. Increase area of natural wetland-upland transition zone to facilitate marsh migration, so that at least 40% of the wetland perimeter is bounded by transition zone that extends inland for at least the full estimated tidal extent under 24 inches (0.6 m) of sea level rise.

C. Increase areas of natural wetland-upland transition zone up to 1,600 feet (500 m) from the marsh edge (depending on topography), even in areas that are not contiguous with the marsh.

D. If the system has a river or creek , then an additional focus should be the creation of adjacent habitats that will allow the migration of wetlands upstream, at least to the head of tide under 24 inches (0.6 m) of sea level rise.

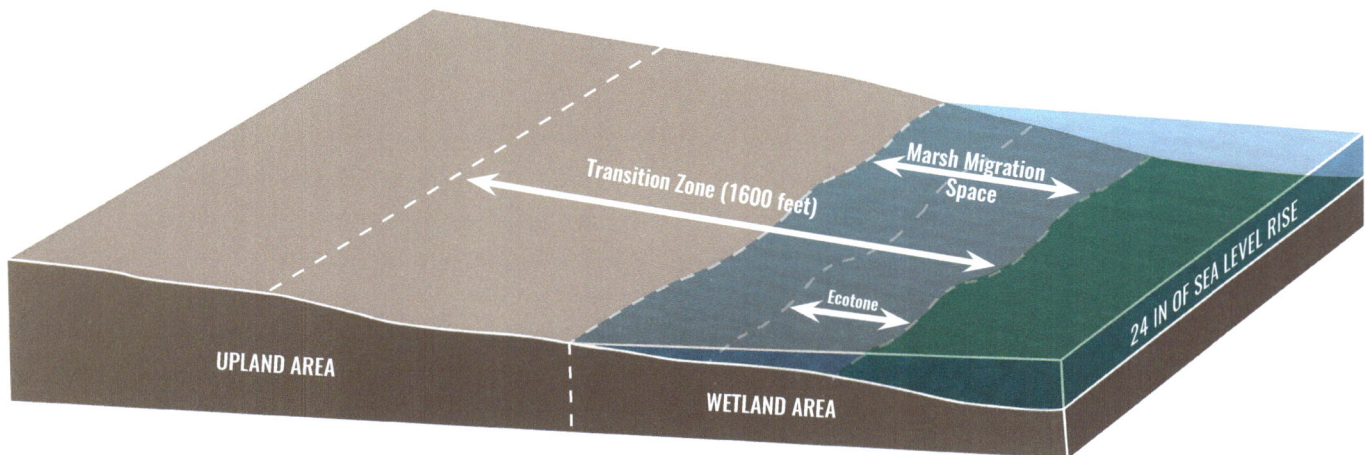

Figure 22. Conceptual diagram depicting wetland area, ecotone, upland area, marsh migration zone and transition zone.

Rationale:

A. Protect all existing natural areas of wetland-upland transition zones from the wetland boundary/edge out to 1,600 feet (500 meters).

Wetland-upland transition zones are areas that often attract development and much of this area has been lost (Appendix 9). Transition zone habitats are important for native wildlife populations, many of which are special status species in coastal Southern California. Protecting existing transition zone adjacent to existing wetlands enhances physical processes that make the shoreline resilient and biological processes that support healthy native wildlife populations. Gradual wetland-upland transition zones allow marsh animals, particularly small mammals, to escape flood waters and reduce wave heights during storms thus reducing erosion and coastal flooding. The proposed wetland-upland transition zone width of 1,600 feet (500 m) was developed through a literature review (Appendix 9).

B. Increase area of natural wetland-upland transition zone to facilitate marsh migration, so that at least 40% of the wetland perimeter is bounded by transition zone that extends inland for at least the full estimated tidal extent under 24 inches (0.6 meter) of sea level rise.

Over time, the land within the 24 inch (0.6 m) sea-level rise elevation band should be made available and accessible for marsh migration. Given the strong likelihood that sea-level rise will eventually go beyond 24 inches (0.6 m), additional inland area that would accommodate even higher sea level rise should be anticipated and built into planning.

The value of 40% represents the average proportion of the existing wetland perimeter, regardless of width, that is currently undeveloped and could potentially become transition zone, for all the wetlands considered by the WRP. Therefore, 40% may not be achievable in some wetlands. Much of this area, although undeveloped, may not be in public ownership, or it may be managed open space, such as a park. It may also require restoration actions, such as removing levees, to achieve the full range of functions, especially marsh migration.

C. Increase areas of natural wetland-upland transition zone up to 1,600 feet (500 meters) from the marsh edge (depending on topography, Appendix 9), even in areas that are not contiguous with the marsh.

Natural areas higher than the marsh migration zone provide connectivity for wildlife in many ways: between wetlands along the coast, between coastal wetlands and river valleys, and between different areas of transition zone itself. Broader transition zone areas with natural physical processes from tides, streams, and hillslopes create gradients of salinity, moisture, and plant communities that promote wildlife diversity at varying distances from the wetlands. Connectivity is important for the persistence of populations, especially in small habitat patches, by allowing refuge from high water, movement between habitat patches, and gene flow. Open space areas and habitat patches of any size throughout the landscape serve as stepping stones and seed sources for colonization.

D. If the system has a river or creek, then an additional focus should be the creation of adjacent habitats that will allow the migration of wetlands upstream, at least to the head of tide under 24 inches (0.6 meter) of sea level rise.

Many of the former upland areas around coastal wetlands have been restricted by development. In many cases the best opportunity for creating and preserving space for marsh migration is along river or creek valleys, taking advantage of the rising topography. Allowing for marsh migration upstream will also improve movement corridors for species between the coastal wetlands and their associated watersheds.

Management Strategies To achieve Objective 5, all of the Management Strategies for Objective 1 should be utilized. Additionally, Objective 5 expands the area of interest by aiming to preserve and restore land beyond the migration zone. Within the migration zone, habitats can include riparian forest, non-tidal brackish marsh, valley freshwater marsh, and other estuarine and palustrine habitat. In many cases, the transition between fresh, brackish, and saline habitats would have been gradual rather than abrupt, and would have varied from year to year. Beyond the migration zone, within the wetland-upland transition zone, habitats can include many upland habitats such as coastal sage scrub, chaparral, grasslands, and woodlands. This larger, wetland-upland transition zone provides ecosystem services including wildlife support (high tide refugia, migration corridors and roosting areas), flood detention areas in floodplains, sequestration, and public access.

Management Strategies 1, 2, 3, and 4 (see Objective 1), and the following:

> **Management Strategy 10: Protect, manage and acquire adjacent land within the wetland-upland transition zone.** Areas that may be suitable as wetland-upland transition zones are not necessarily in public ownership and may be subject to development pressures, making land acquisition in transition zones a challenge. Protecting adjacent open space either by acquisition or by easement should be a priority. While Management Strategy 2 focuses on adjacent land for wetland migration, this Management Strategy is moving beyond the migration zone. Meaning the adjacent land needed for Objectives 5A and 5C, is even further inland to the area of land inundated by 24 inches (0.6 m) of sea-level rise. This adjacent land does not necessarily need to be contiguous. While the existence of some structures within the wetland-upland transition zone may be okay, those structures should not impeded wildlife movement. These structures should also be potentially removable in the future when more land is needed for migration due to an increase in sea-level rise beyond 24 inches (0.6 m).

Objective Tracking: Create a habitat map of transition zones for the project area (developed above for Objective 4, Habitat Diversity). For all transition zone Objectives, estimate the existing and proposed (project) transition zone width as well as the existing and proposed percent of the tidal wetland boundary that is adjacent to the transition zone. Project should further determine how many acres of additional transition zone habitats will be protected that are outside the project boundary. See Appendix 9 for transition zone mapping methodology.

WETLAND-UPLAND TRANSITION ZONES

The area adjacent to wetlands has many functions and has many different names (Table 6). The term "buffer" describes the land between landscape stressors and wetlands, and is used in a variety of regulatory contexts. A buffer area is used to protect wetlands from stressors and is not defined in terms of ecological functions and services. For the purposes of this document, the term buffer is used for non-tidal wetlands in Goal 2, whereas three additional terms are used to describe the areas adjacent to tidal wetlands (Figure 22). The wetland-upland transition zone includes non-tidal habitats adjacent to the coastal wetland edge, up to 1,600 feet (500 m) wide, that encompass the ecosystem functions and services associated with the wetlands, and can include habitats such as alkali wetlands, riparian areas, coastal sage scrub, and many other upland habitats. Similarly, the wetland-upland

TRANSITION ZONE IN LOS PEÑASQUITO • PHOTO COURTESY OF JEREMY LOWE

DEVEREAUX SLOUGH • PHOTO COURTESY OF USFWS

ecotone is a narrow band of habitat where wetlands and uplands meet, and contains vegetations types from both habitats (Figure 22). The ecotone boundaries are set by factors such as soil salinity and moisture (Callaway et al. 1990; James and Zedler 2000). The sea-level rise migration zone is a particular zone, identified by land elevation, that will accommodate the wetland's upslope movement as sea-levels rise. The wetland-upland transition zone includes both the narrow ecotone and sea-level rise migration zone, as well as further inland adjacent habitats (Figure 22). The transition zone provides important refuge for marsh wildlife, and allows upland wildlife to access the marsh for food and other resources. These areas support gradients in environmental variables such as salinity, soil moisture, and temperature that can be important to supporting adaptation within wildlife populations, and can also support unique habitat types (e.g. alkali wetlands, salt pannes) that further contribute to landscape complexity.

Much of the historical transition zone habitat in California has been lost due to competing land uses along the shoreline. Accelerating sea-level rise increases the challenge of supporting transition zone habitats, and ecosystem services associated with transition zones, especially in heavily developed areas. Protection and restoration of the wetland-upland transition zones are critically important if tidal wetlands and their functions are going to persist.

The transition zone varies by location. Gradual hillslope transitions provide opportunities to support wide habitat gradients, biological diversity, and landscape complexity. Extending the upper boundary for potential transition zone 1,600 feet (500 m) from the lower boundary of the transition zone captures the majority of these biological diversity support functions. For cliff and bluff transition zones, the area at the top of the bluff is unlikely to provide the same flood control, habitat gradient, and movement corridor benefits as hillslope transition zones. Riverine transition zones transition between fluvial and tidal processes and conditions and are important for floodwater storage and retention, as well as supporting a unique assemblage of plant and wildlife species.

Land use is a major consideration in determining what can be done within the transition zone. Although many developed areas are unlikely to be considered for potential restoration, these areas may still support transition zone functions in other ways. For example, some land uses in developed areas (e.g., vacant lots, golf courses) may still provide some buffering functions. Actions taken in developed areas within this boundary can support wildlife movement (e.g., removal of barriers and planting of native vegetation in yards), and affect flood control (e.g., rain gardens and bioswales).

See Appendix 9 for more information wetland-upland transition zones.

Transition Zone Terms	WRP Definition
Buffer (for streams, adjacent habitats, and other non-tidal wetlands)	Native habitat beyond the riparian zone (i.e., "adjacent habitats") ranging in width from 100 to 3,000 feet (30-100 m) that protect the stream, adjacent habitats, and/or other non-tidal wetlands from the impacts of adjacent land-uses
Wetland-upland transition zone	Non-tidal habitats adjacent to the coastal wetland edge, up to 1,600 feet (490 m) wide
Wetland-upland ecotone	Narrow band of habitat between wetlands and uplands
Sea-level rise migration zone	Land required to accomodate marsh migration with sea-level rise

A. Restore tidal characteristics (range, extent and residence time) guided by appropriate reference conditions, to support habitat abundance and distribution as indicated in Objectives 1–4.

B. Restore freshwater and sediment flow characteristics from watersheds (volume, frequency, and timing) guided by appropriate reference conditions, to support habitat abundance and distribution as indicated in Objectives 1–4.

C. Restore or manage sediment inputs to maintain wetland and transition zone elevations sufficient to accommodate 24 inch (0.6 m) of estimated sea level rise. Inputs should be assessed based on total annual volume and magnitude of peak inputs.

Rationale:

A. Restore tidal characteristics (range, extent, and residence time) guided by appropriate reference conditions, to support habitat abundance and distribution as indicated in Objectives 1–4.

The intent of this Objective is to ensure that 100% of coastal wetlands are hydrologically connected with the ocean at periodicities and magnitudes similar to appropriate reference systems conditions. An appropriate reference condition is the best representation of a target habitat and/or ecosystem process existing within the same subregion (see Monitoring Section on page 77). The tidal inlet state will have significant impacts on the water quality of the estuary, dependent upon the tidal state of the estuary when closure occurred and the amount of freshwater dilution from the river. In many cases the tidal inlet has been modified by the construction of jetties or by a regime of managed breaching. Also, hydrologic and ecologic connections in many wetlands have been impeded by the construction of dams, berms, and by draining and filling, or altered by water importation and by increased dry weather runoff from urban areas. Groundwater levels have also been affected by pumping to a degree that, in places, natural recharge cannot occur. This Objective aims to maintain and improve hydrologic and ecologic connectivity of coastal wetland systems including their connection with the ocean, their associated watersheds, and groundwater basins. For more information on selecting a reference condition, see page 79.

B. Restore freshwater and sediment flow characteristics from watersheds (volume, frequency, and timing) guided by appropriate reference conditions, to support habitat abundance and distribution as indicated in Objectives 1–4.

This Objective aims to maintain and improve hydrologic and ecological connectivity of coastal wetland systems with their connection to their associated watersheds. The intent is to ensure that 100% of coastal wetlands are hydrologically connected with their associated watersheds at periodicities and magnitudes similar to appropriate reference systems conditions. This includes management for increased or decreased watershed inputs, and addresses both water and sediment inputs.

C. Restore or manage sediment inputs to maintain wetland and transition zone elevations sufficient to accommodate 24 inch (0.6 m) of estimated sea level rise. Inputs should be assessed based on total annual volume and magnitude of peak inputs.

This Objective aims to provide optimal wetland resilience in the face of sea-level rise by reconnecting coastal wetlands to the appropriate watershed and oceanic sediment supplies. With a direct connection to a sediment source, coastal wetlands are able to accrete sediment on the marsh surface and potentially gain enough elevation to keep pace with current and future rates of sea-level rise.

Management Strategies Management Strategies 5–7 (Objective 3) should also be employed for Objective 6. In addition, the WRP recommends three additional Management Strategies focused on reconnecting sediment supply to coastal wetlands (Figure 23).

For some coastal wetlands that will have limited migration space, like small creeks and small lagoons, increased rates of sediment accretion might be the only strategy that will enable them to persist. Increasing sediment accretion can occur by artificial augmentation (e.g., spraying dredge material) or by restoring natural sources, such as removing a dam.

There are a few coastal wetlands in Southern California that currently experience an excess of sediment input from their watersheds (e.g., Tijuana River Estuary and Los Peñasquitos Lagoon). While this large sediment load has been a burden in the past, especially during large storm events, it is important to recognize that management of sediment and water in these few watersheds may need to change over time. While large sediment pulses may be a nuisance today, that same material may be critical in the future to reduce the impacts of flooding associated with sea-level rise. For instance, management actions such as building detention basins and levees to keep sediment out may need to stop in the future, or at the very least, the sediment that is retained by those actions should be stored for future use within the system (e.g., for sediment augmentation or for building wetlands in the nearshore). This future shift in management actions also applies to dredging operations for channel navigation in Southern California's larger bays. Maintenance dredging operations should work with restoration managers to increase beneficial reuse of dredged material in order to help coastal wetlands keep pace with sea-level rise.

Management Strategies 5, 6, and 7 (see Objective 3), and the following:

Management Strategy 11: Remove barriers to release sediment held higher in the watershed. The presence of dams, detention basins and other structures higher up in the watershed often inhibit the flow of water and sediment to lower reaches. The removal of such structures allows sediment and water to move in a more natural manner down the creek to the wetlands. However, there needs to be a commensurate change in management for flooding and sediment accretion in the channels (Management Strategy 2).

Management Strategy 12: Manage flows in river channels to increase their capacity to move sediment from the watershed. Hydrological flows and sediment supply need to be balanced so that the river channel has sufficient capacity and competence to move sediment downstream through the watershed to the tidal wetlands. This may be achieved by modifying the timing and volume of releases of water detained higher up in the watershed or through altered maintenance of debris basins. Mechanical methods may be necessary to move sediment through the watershed into the estuaries if natural transport processes are impractical.

Management Strategy 13: Augment sediment processes to raise and maintain marsh elevation. In some instances, berms almost completely surround and cut off hydrological and sediment flows between creeks and wetlands (e.g., Seal Beach National Wildlife Refuge). In this situation, it may be necessary to artificially augment sediment processes to raise and maintain marsh elevations. One example of this Management Strategy is the use of thin layer placement by pumping slurry or spraying to facilitate desirable marsh elevations.

Tracking Objectives:

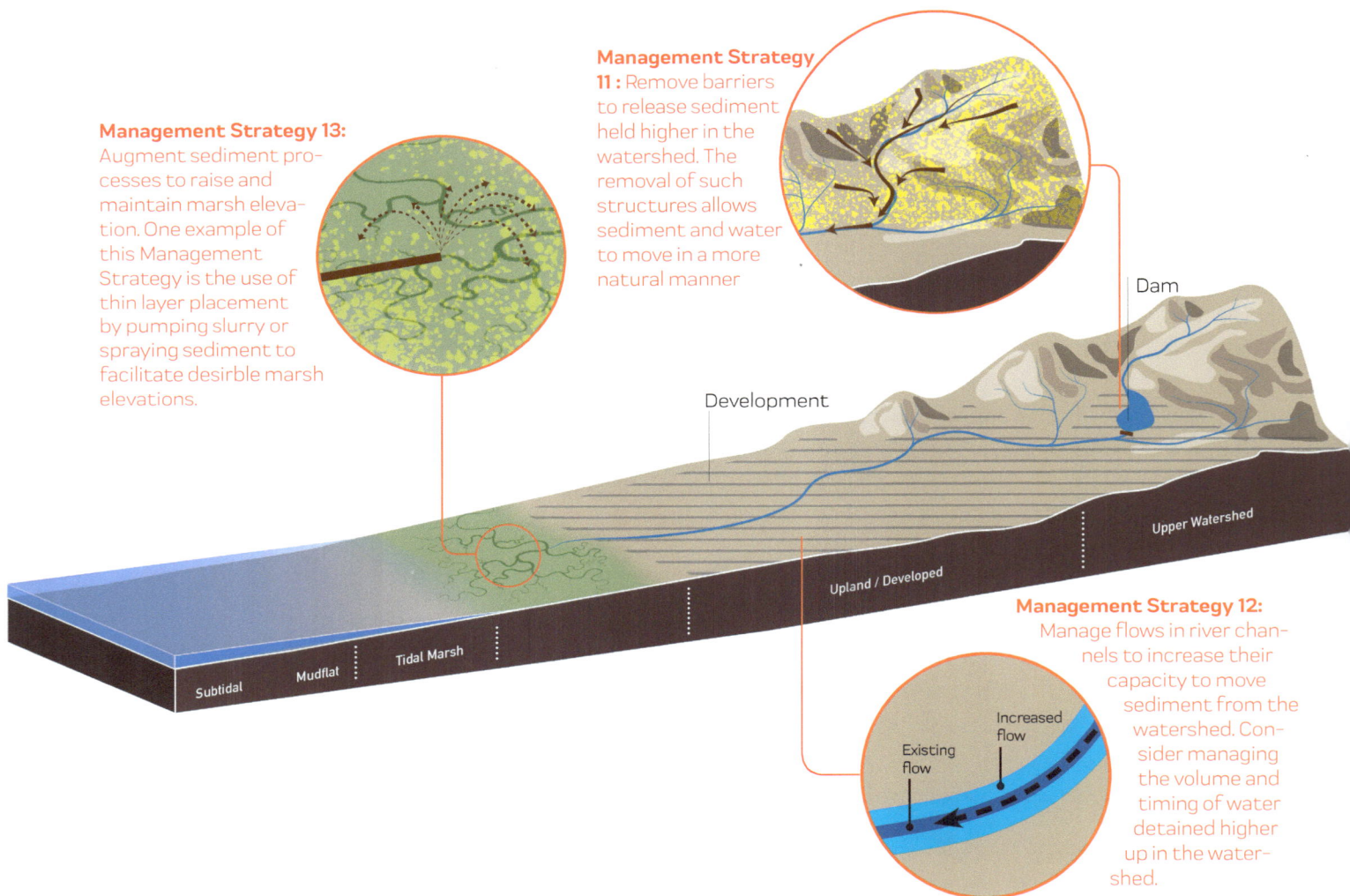

Management Strategy 13: Augment sediment processes to raise and maintain marsh elevation. One example of this Management Strategy is the use of thin layer placement by pumping slurry or spraying sediment to facilitate desirble marsh elevations.

Management Strategy 11 : Remove barriers to release sediment held higher in the watershed. The removal of such structures allows sediment and water to move in a more natural manner

Dam

Development

Upper Watershed

Upland / Developed

Subtidal   Mudflat   Tidal Marsh

Management Strategy 12: Manage flows in river channels to increase their capacity to move sediment from the watershed. Consider managing the volume and timing of water detained higher up in the watershed.

Existing flow

Increased flow

Figure 23. Conceptual diagram showing Management Strategies 11–13.

LIVING SHORELINES

A living shoreline refers to a shoreline with natural elements, such as beaches, marshes, oyster reefs, and eelgrass beds, that is designed to restore habitat as well as help in physically protecting the coast from the effects of large storm events and sea-level rise. This can be alone or in combination with hard structures such as levees and seawalls. Living shorelines could be useful shoreline protection elements in situations where infrastructure is removed for habitat restoration purposes. While living shorelines often rely on the planting of native vegetation, sometimes other less-natural materials such as stone sills, coir logs, oyster reef elements such a "reef balls" or shell material, or grounds are included to further reduce wave energy or trap sediment.

To date, California has implemented native Olympia oyster reefs, eelgrass beds, tidal wetlands revegetation, upland ecotones, sand beaches, and coastal dune restoration projects as living shorelines. The California State Coastal Conservancy has funded several living shoreline projects in Southern California including the Upper Newport Bay Living Shoreline Project (oyster and eelgrass restoration) and the San Diego Bay Native Oyster Restoration Project. Other potential habitats include coastal islands and boulder fields, kelp forests and other types of seaweed beds, rocky intertidal areas, and coastal bluffs as appropriate to the local geomorphology and ecology.

However, Southern California is highly urbanized with one of the most hardened coastlines in all of the United States, and space is limited. Living shoreline projects will need to be coupled with the managed removal and realignment of built infrastructure to provide the space needed for coastal ecosystems to function properly.

See Appendix 10 for more information on living shorelines.

SEDIMENT SPRAYING AT SEAL BEACH NATIONAL WILDLIFE REFUGE • PHOTO COURTESY OF USFWS

The tracking procedure for this Objective will be designed for the specific needs of each project. Each project will use monitoring metrics that address the project Objectives while also integrating them with the project's site-specific monitoring program.

A. Each project's site-specific monitoring program should assess the tidal characteristics, including tidal range, tidal extent, residence time and frequency of inlet opening and closing. Inundation regime should approximate appropriate reference systems.

B. Appropriate levels of water and sediment flow into and out of tidal wetlands will play a determining role in their abilities to adapt to rising sea levels. Projects should model water and sediment levels necessary to maintain desired elevations and duration/extent of inundation for the restoration design. To assess reconnections of tidal wetlands to the ocean and/or watersheds, project monitoring should include water level and turbidity loggers at the connection points for water and sediment, including the tidal inlets and the sources of freshwater. The performance criteria will be evaluated in comparison to a reference condition.

C. Prior to project implementation, pre-construction sea level rise modeling should demonstrate that the site will keep pace with 24 inches (0.6 m) of sea-level rise. See Appendix 11 for more information on selecting sea-level rise and marsh evolution models. Post-construction monitoring of sediment accretion rates will groundtruth the modeling and assess the site's ability to keep pace with 24 inches (0.6 m) of sea-level rise.

KING TIDE, DECEMBER 2016 AT ALAMITOS BAY, LONG BEACH • PHOTO COURTESY OF CREATIVE COMMONS

A. Improve the major attributes of wetland condition, including biology, hydrology, physical structure, and landscape context, as measured by a rapid assessment score, for 100% of systems within each archetype.

B. 100% of mature coastal wetlands (i.e., natural coastal wetland or restored coastal wetland of 40 years or more) should achieve and maintain an overall CRAM score ranging from 76-94.

C. 100% of future restoration projects should be on or above the Habitat Development Curve based on the project age as the restoration matures.

Rationale: The previous Objectives (1–6) have focused on the abundance, system characteristics, and connectivity of coastal wetlands systems. This last Objective focuses on the wetland condition—wetland status now and in the future to provide the expected functions and habitats to support aquatic and wildlife species for the same type of wetland in its natural, undisturbed, setting. Condition means the status of physical, chemical, biological, and ecological indicators of the levels of services and beneficial uses of wetland systems. For this Objective, condition is recommended to be assessed using the indicators defined by the California Rapid Assessment Method (CRAM) based on the California Wetland and Riparian Area Monitoring Plan (WRAMP) approach (CWQMC 2016). CRAM is a cost-effective and scientifically-defensible method for monitoring wetland condition.

A. Improve the major attributes of wetland condition, including biology, hydrology, physical structure, and landscape context, as measured by a rapid assessment score, for 100% of systems within each archetype.

The condition profile for the region, as shown by the Cumulative Distribution Function (CDF) (Figure 24), can be used to set performance criteria for projects at the regional scale. (CWMW 2008; Collins and Stein 2018). Unless new projects score above the 50th percentile score for the region, at a CRAM score of 65, they degrade the region's condition profile. To improve the condition profile, new projects should score above the 50th percentile. Higher scores for larger projects will improve the profile more because they represent more of the wetland resource.

To develop a condition profile, project scores are plotted on the CDF for the wetland type. The CDF can be developed using existing appropriate CRAM scores taken from the eCRAM database.

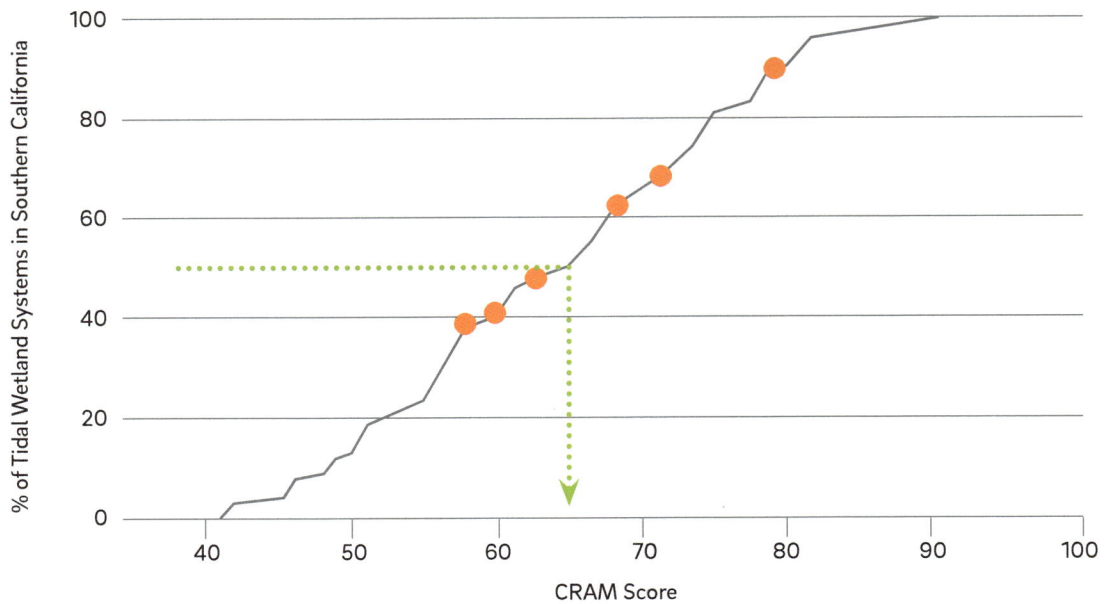

Figure 24. Cumulative Distribution Function (CDF) of the overall condition of coastal wetlands in Southern California derived from a 2008 probabilistic ambient survey using CRAM, showing a 50th percentile CRAM score of 65. Orange dots represent projects.

B. 100% of mature coastal wetlands (i.e. natural coastal wetland or restored coastal wetland of 40 years or more) must achieve and maintain an overall CRAM score ranging from 76–94 (i.e., the reference envelope shown in Figure 25).

While restored wetlands evolve over time, ecosystem function should mature a few decades after restoration activities occur. For instance, in Southern California restored wetlands typically score within the reference envelope (grey band in Figure 25) ~40 years of age. The reference envelope represents the range of CRAM scores natural wetlands have scored. In Southern California, the reference envelope ranges in CRAM scores from 76–94. As indicated by the black diamonds on Figure 25, Southern California coastal wetland projects tend to be younger and score lower than other wetlands throughout California. This Objective aims to increase Southern California's natural and restored coastal wetlands' CRAM scores once they have reached maturity.

C. 100% of restoration projects must be on or above the Habitat Development Curve based on the project age as the restoration matures.

Restored wetlands will evolve over time, and the wetland condition should be improving with age. The trajectory of improvement can be assessed using Habitat Development Curves (HDCs) based on CRAM (CWMW 2008) (Figure 25). HDCs are produced by plotting the wetland condition of many systems against wetland age and reference condition. When the HDC is based on CRAM, it quantifies the rate of habitat development as the increase in CRAM scores over times. HDCs exist for coastal, riverine, and depressional wetlands.

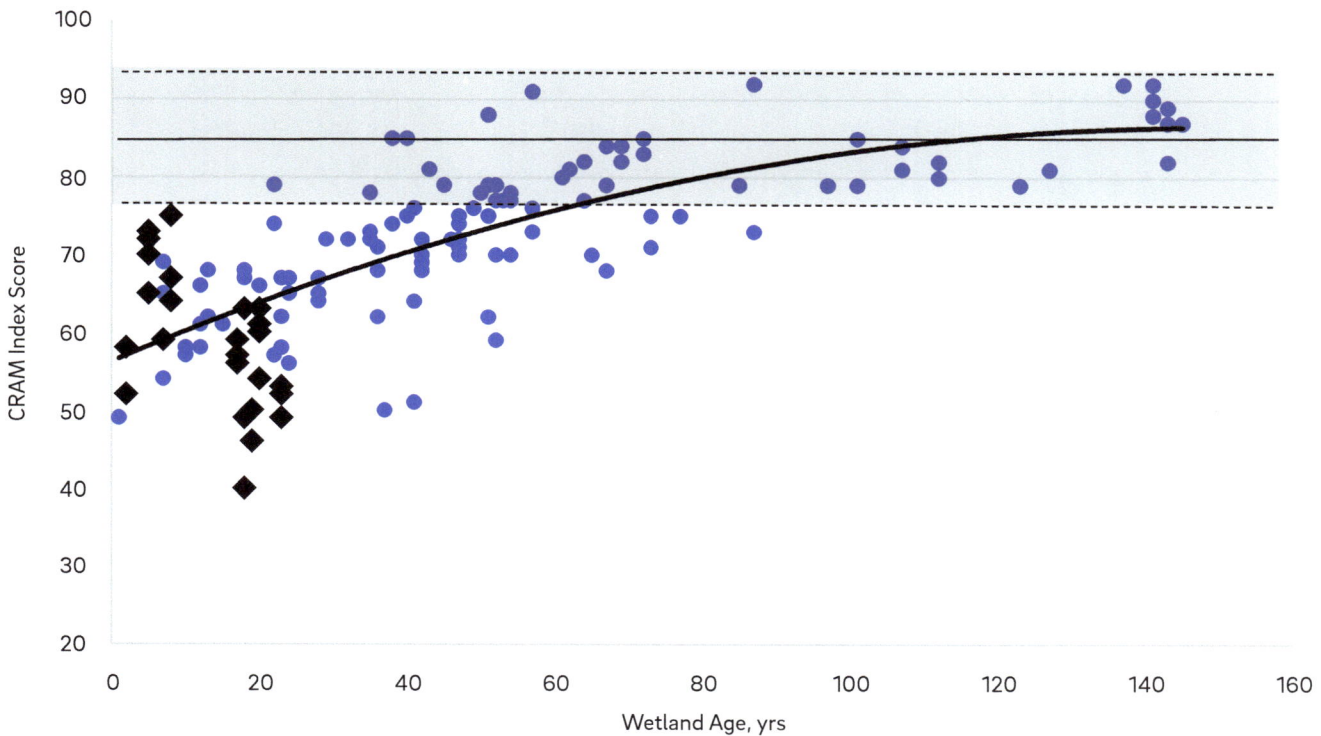

Figure 25. Habitat Development Curve (HDC) for coastal wetlands of California based on a 2008 probabilistic survey of natural wetlands and projects. Grey area represents the reference envelope based on CRAM scores in Southern California reference sites. Blue dots represent natural sites across California and black diamonds represent coastal wetland restoration projects in Southern California.

Management Strategies:

Management Strategy 14: Conduct pre-construction CRAM assessment and project anticipated CRAM score for site when mature.

Management Strategy 15: Review post-construction CRAM score and compare to project's evolution to the Habitat Development Curve.

Tracking Objectives: Pre-construction and post-construction CRAM scores should be documented for each project. CRAM scores taken at various times after project implementation should be compared with the projected Habitat Development Curve. Monitoring will help the WRP assess trends in wetland extent and condition and relate these trends to management actions, climate change, and other natural and anthropogenic factors in a way that informs planning and management decisions throughout the State (see Recommendations for a Comprehensive Regional Monitoring Program for Coastal Wetlands on page 77).

TIMELINE TO ACHIEVE OBJECTIVES FOR GOAL 1

Implementing the Objectives and Management Strategies will involve the design and implementation of new and innovative projects. Complex and/or large-scale coastal wetland restoration projects in Southern California take 10 to 20 years from initial planning to implementation. Once implemented, it may take several additional decades for the wetland to evolve into a fully-functioning wetland ecosystem (Figure 26). As demonstrated in the habitat change analysis on pages 24–29 , Southern California will lose an additional 800 acres (320 ha) of coastal wetlands after 24 inches (0.6 m) of sea-level rise if restoration actions are not taken immediately. Due to this potential loss, it is critical that coastal wetland restoration projects are started as soon as possible in order for these systems to establish resilient and properly functioning ecosystems before they face constant inundation from rising tides.

The most recent sea-level rise projections suggest that the rate of sea-level rise will increase markedly after 2050, resulting in a near doubling of the rate by 2100 (Griggs et al. 2017). Considering that restoration can take up to 20 years after completion to achieve a fully functioning coastal wetland, restoration projects should be completed by 2030 in order to establish mature marshes by 2050 (Figure 27). If the ultimate aim is to establish resilient vegetated marshes by 2050, such planning needs to begin immediately to be realized within this timeframe.

Figure 26. Typical time required to plan, design, and implement a wetland restoration project and for the evolution toward a functioning wetland.

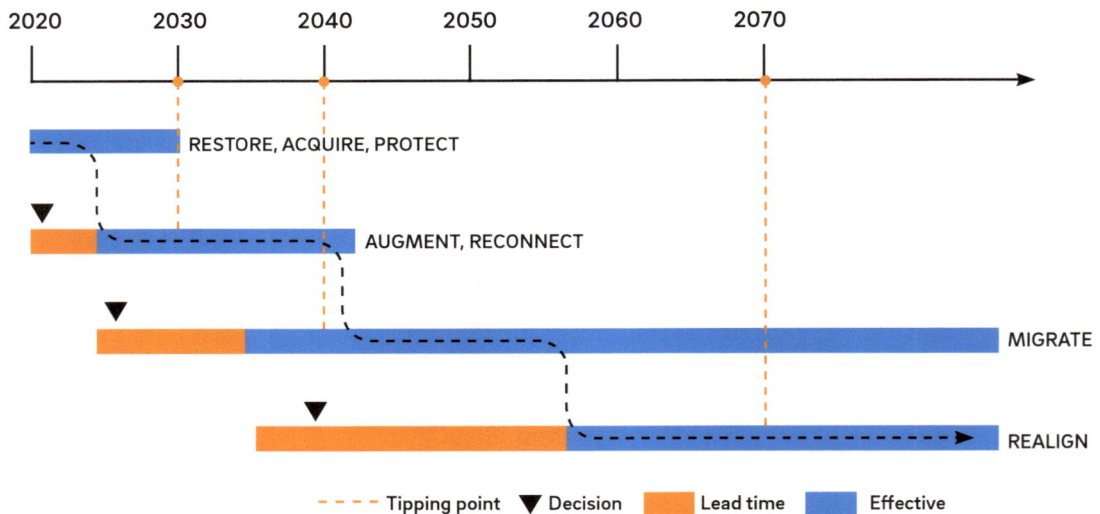

Figure 27. Prioritized Management Strategies over time.

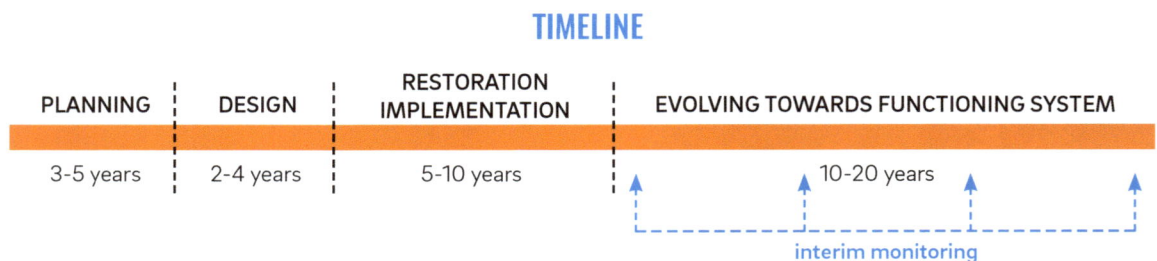

It is also important to anticipate other challenges that may accompany rising sea levels and, thereby, pose a challenge to implementing Management Strategies. For example, the 2020 to 2030 timeframe will also be a time of increasing pressure to armor the shoreline as the effects of sea-level rise on flooding become more apparent. This will, in turn, increase pressure on the wetlands and make establishing resilient functioning marshes by 2050 even more of a challenge. With these challenges in mind, Figure 26 also identifies tipping points where if action is not taken, it may be too late to mitigate the impacts of sea-level rise. For instance, by 2040 scientists expect many wetlands will start becoming "squeezed" and by 2070 many vegetated marsh plains will start to drown, converting to unvegetated flats and shallow subtidal areas (Stralberg et al. 2011).

Not every wetland project will need to undertake every Management Strategy and even if several are required, they do not need to be done simultaneously. Management Strategies should be prioritized for each project to identify the activities that need to be undertaken sooner rather than later (Table 7).

Table 7. Suggested timing for implementation of management strategies for coastal wetlands.

Strategy Group	Management Strategies in Order of Priority
Acquire Wetlands and Transition Zones	Protect, manage and acquire adjacent land within the wetland migration zone (Management Strategy 2).
	Protect, manage and acquire adjacent land within the wetland-upland transition zone (Management Strategy 10).
Reconnect Hydrological and Sediment Processes	Allow tidal inlets to open and close naturally (Management Strategy 6).
	Modify or remove structures to restore inundation regime (Management Strategy 7).
	Remove barriers to release sediment held higher in the watershed (Management Strategy 11).
	Manage flows in river channels to increase their capacity to move sediment from the watershed (Management Strategy 12).
	Augment sediment processes to raise and maintain vegetated marsh (Management Strategy 13).
Remove Barriers to Wetland Migration	Remove barriers that prevent wetlands from expanding or migrating (Management Strategy 1).
	Remove barriers to reconnect channels to wetlands (Management Strategy 5).
Realign Barriers to Wetland Migration	Grade areas adjacent to wetlands to increase opportunity for migration (Management Strategy 3).
	Relocate or modify adjacent infrastructure or development (Management Strategy 4).

SOONER

LATER

Acquire Wetlands and Transition Zones (Strategies 2 and 10): Strategies to protect and acquire land not in public ownership, particularly open adjacent upland areas that are or could become transition zone habitat, should be heavily prioritized since these areas are subject to high development pressures and may not be available in the future. The availability of wetland migration areas within wetland-upland transition zones will determine the fate of wetlands in the future. Wetlands with no available migration space will be converted to subtidal habitats. Opportunities for habitat creation in the wetland-upland transition zone will always be of value regardless of its direct connection to the wetland since this habitat is rare in Southern California and will provide immediate benefit to the estuary as well as longer term benefits for sea-level rise adaptation.

Reconnect Hydrological and Sediment Processes (Strategies 6, 7, 11, 12 and 13): Strategies to accelerate or augment the vertical accretion of marshes may be necessary now, if sediment supplies are low, or become necessary in the near future as sea-level rise accelerates. This could involve the reconnection of creeks or channels to wetlands and, in some cases, direct application of sediment into wetlands (e.g., thin layer placement). A wetland's ability to keep pace with sea-level rise within the existing footprint will be determined by the sediment supply and associated accretion rates.

Remove and Realign Barriers to Wetland Migration (Strategies 1, 3, 4 and 5): In the longer-term, the removal or realignment of barriers to wetland migration, such as berms, will be necessary. This may require the relocation of infrastructure and developments through managed retreat and would therefore require a longer, more involved, planning process. Projects to removing barriers should begin before realignment because they will be more complicated projects and barriers that prevent migration need to be removed before wetlands will have space for retreat.

Strategies 8 and 9 (not listed in table): Habitat restoration activities will occur in conjunction with, and throughout, implementation of the strategies to improve physical processes listed in Table 7. Protecting habitats like salt flats and subtidal should be high priorities within restoration projects.

Successful implementation of prioritized strategies will require working closely with regulatory and funding agencies to allow for new and creative solutions for protecting and restoring wetlands into the future. Funding and permitting wetlands restoration projects that include upland areas will be a new concept for many agencies, but we need to start these new practices now. Acquiring and protecting upland areas and transition zones today must be done in order to ensure that coastal wetlands exist in Southern California in the future.

RECOMMENDATIONS FOR A COMPREHENSIVE REGIONAL MONITORING PROGRAM FOR COASTAL WETLANDS

Wetland monitoring, reporting, and data-sharing are the best tools we have to evaluate the success of the WRP's efforts. While we can track progress toward our quantitative Objectives by assessing individual projects, project-tracking does not assess the impact of the *Regional Strategy 2018* on all coastal wetlands throughout the region. For that we need a comprehensive monitoring program. A comprehensive monitoring program would assess the effectiveness of the collective actions of the WRP and others on the abundance, diversity, and condition of wetlands throughout the region. Such a program would also help identify needs for adaptive management, revise Objectives as necessary based on changed conditions or new knowledge, and learn from past experiences to help improve future restoration and management efforts. To fulfill these purposes, a monitoring program would need to assess the individual and collective effects of wetland-related projects throughout the Region. This section provides a recommendation for development of a comprehensive wetland monitoring program.

The WRP has been involved with supporting and developing regional monitoring tools for almost twenty years. The Integrated Regional Wetlands Assessment Program (IWRAP) (http://www.irwm.org/) was developed in 2007 to provide a framework for regional wetlands monitoring (CWMW 2008). The most recent work is a report entitled Regional Monitoring Report for Southern California Coastal Wetlands: Application of the USEPA Three-Tiered Monitoring Strategy (Johnston et al. 2015). These past efforts should be built upon as regional monitoring moves forward. The WRP is well positioned to provide the regional coordination needed to implement a multi-agency regional monitoring program.

Monitoring Framework

A regional monitoring program for coastal wetlands should be consistent with the Wetland and Riparian Area Monitoring Plan (WRAMP) developed by the California Wetland Monitoring Workgroup (CWMW 2010), and should build off previous efforts. The WRAMP is a framework and toolset developed by statewide technical teams, with oversight by the CWMW, based on the three categories of wetland monitoring data defined by the USEPA. The WRAMP toolset includes standard methods for developing the three categories of data below:

- **Level 1: Map-based inventories** that answer questions about wetland abundance, distribution, and diversity. The primary Level 1 tool is the California Aquatic Resource Inventory (CARI).

- **Level 2: Rapid, field-based assessments** of overall wetland condition and stress, relative to the best achievable conditions. The primary Level 2 tool is the California Rapid Assessment Method (CRAM).

- **Level 3: Quantitative field-based measures** of specific aspects of wetland condition or stress. The WRP has sponsored the development of Level 3 protocols for monitoring tidal marsh (Johnston et al. 2015). In addition, there are protocols for Santa Monica Bay (Johnston et al. 2015), the San Francisco Bay Wetlands Regional Monitoring Program (CWMW 2010), the Interagency Ecological Program (IEP) (Sacramento-San Joaquin Delta; IEP 2012), and three National Estuarine Research Reserves in California (Tijuana Estuary, Elkhorn Slough, San Francisco Bay; Kennish 2004).

A comprehensive monitoring program should utilize the WRAMP Framework and existing Level 1-2-3 tools as much as possible (https://mywaterquality.ca.gov/monitoring_council/wetland_workgroup/wramp/index.html). The following is the proposed structure for the WRP's regional monitoring program.

Level 1: Regional Monitoring. The most fundamental component the monitoring program is a comprehensive inventory of coastal wetland habitats. The inventory allows for evaluation of overall areal changes and the influence of projects, and serves as the sample frame for assessment of overall condition. The habitat inventory is essential for assessing the distribution, abundance, and diversity of wetland habitats. The existing inventory developed by SCCWRP in 2008 (https://www.csun.edu/center-for-geographical-studies/scwmp-project-data-and-reports) should be updated as a regional version of the statewide California Aquatic Resource Inventory (CARI). The Southern California coastal version of CARI (SCCARI) should adopt the regional habitat nomenclature and typology developed for the WRP. It should serve as the base map for displaying WRP projects and other habitat projects in the California EcoAtlas information system (CWMW 2018), with linkages to the WRP Decision Support System (scwrp.databasin.org). Ongoing updates of SCCARI should be achieved through local and regional agencies using the online CARI editor, adapted to SCCARI. Regional remapping is generally not necessary, if significant spatial changes in habitat are captured through ongoing updates done by local and regional agencies as routine aspects of their projects.

Level 2: Rapid Assessment. While each WRP project's pre- and post-construction California Rapid Assessment Method (CRAM) surveys will significantly contribute to understanding wetland function at a regional scale, the most cost-effective way to survey the regional and subregional wetland conditions is to assess them through a probabilistic survey design using CRAM. Probabilistic surveys are most cost-effective for assessing regional conditions with adequate statistical confidence. A regional probabilistic survey of tidal marsh condition was conducted by SCCWRP in 2008 (Sutula et al. 2008) and could serve as the baseline measure of condition for the purpose of tracking the effects of WRP projects regional conditions going forward. Regional surveys of conditions might also include Level 3 data, if they are necessary and affordable. Level 3 data commonly considered in regional habitat surveys relate to the distribution and abundance of plants and wildlife, especially species of special status, and the levels of habitat invasion and chemical contamination. There is no set interval for regional resurveys of wetland condition. In general, the interval depends on the expected rate of change in condition due to climate change, land use change, amounts of habitat restoration, or the advent of extreme environmental events, such as major sudden pollution, catastrophic wildfires, or catastrophic floods.

Level 3: Project Monitoring. Every wetland restoration project will need to be monitored to comply with permits, assess its contribution to progress toward WRP regional Goals, and to account for public investments in habitat restoration. Project monitoring should include pre-construction, construction, and post-construction monitoring. It will not be necessary, however, to replicate all monitoring for all projects. The monitoring must be carefully designed to assess progress toward the project-specific goals, Objectives, and performance measures, which will vary among projects. New Level 3 protocols may be needed to provide some necessary data. As explained by the WRAMP framework, any Level 3 data must be thoroughly justified, given their relatively high cost and local site-specificity.

Project monitoring can also help assess progress toward the Goals and Objectives of the WRP. The challenge is to focus on indicators that serve both project-specific and WRP monitoring needs. To the extent appropriate and affordable, individual projects should incorporate *Regional Strategy 2018* Objective-tracking measures into their monitoring programs.

- **Sentinel sites**—The regional monitoring program should include routine monitoring at a set of strategically-located sentinel sites. Sentinel sites are wetlands that are designated for long-term monitoring in order to evaluate regional trends in external conditions, such as sea-level rise, weather, large-scale biological changes, or land use change. They can also be useful for evaluating long term efficacy of restoration and management actions. Sentinel sites are important because external conditions can shift baseline conditions and thus influence the interpretation of Level 1 and Level 2 monitoring data. Not all factors have to be monitored at all of the sentinel sites, and they do not need intensive annual monitoring. Sentinel site should be assessed at certain intervals (e.g., every 5 years) or following specific trigger events (e.g., following an El Niño), based on the condition being tracked. Sentinel sites should represent all the wetland archetypes and should include the following categories:

- **Reference sites**—sites that reflect the least altered wetlands in the landscape, and often the sites used to compare reference conditions for project-specific monitoring. A system of reference sites should be developed that can be used for tracking project success. The first step in developing a reference system would be identifying for each archetype what are the key attributes that constitute an appropriately-functioning wetland.

  - **Reference Condition**—Reference condition provides a basis of comparison against which to judge the performance of a restoration site. Depending on the project, the reference condition for a given variable can be anything from the pre-European condition to the best attainable condition given current constraints of the landscape. This is a management decision that is made for each project. Reference conditions can often be monitored at reference sites, but not always because reference sites seldom meet all reference condition; this is particularly true in urban settings where the legacy of past and current land use practices continues to affect the condition of all wetlands.

- **Past restoration sites**—a subset of restoration sites that are tracked over time to understand their long-term ecological progression;

- **At-risk sites**—sites that are identified to be at risk of impact from factors like sea-level rise or a major development project; and

- **Management sites**—similar to restoration sites, but are sites that have been subject to some form of experimental management action (e.g. living shorelines).

A key component of the sentinel site monitoring program should be dedicated tide stations established throughout the region according to NOAA standards for assessing sea-level rise relative to wetland elevation. It is essential to be able to forecast loss of wetlands due to sea level rise early enough to implement mediating actions and to adjust the WRP Goals and Objectives if necessary. The monitoring program should also compile any maps of risks and vulnerabilities due to sea-level rise or other aspects of climate change.

Implementation

The State of California and U.S. Environmental Protection Agency both call for coordinated wetland monitoring to adequately account for the millions of public dollars spent on wetland conservation and restoration projects in California (California Wetland Interagency Team 2017). The regional monitoring program, as recommended here, will meet this obligation. The following steps must be completed to establish this program:

- Develop a monitoring program plan that includes a charter and addresses governance, institutional roles and responsibilities, programmatic relations to other monitoring programs, public reporting, the annual budget, and a business plan;

- Identify a public agency or other organization to administer the program;

- Develop or adapt a regional data management system, with adequate quality control measures, data templates, and online visualization, that is used to manage project-specific, rapid assessment, and sentinel site monitoring data;

- Establish a monitoring workgroup of the WRP to advise and review the implementation of the technical recommendations above and below, focusing on the Objectives of Goals 1 and 2.

SHOREBIRD AT NEWPORT BAY • PHOTO BY TRACIE HALL, COURTESY OF CREATIVE COMMONS

The regional monitoring program will need to adhere to a set of basic operational practices to secure and safeguard its scientific reputation. These practices include, but are not necessarily limited to:

- WRP Work Plan project proponents should utilize the WRP Regional Planning Atlas (scwrp.databasin.org) to compile and share project information and data (only for all projects on the WRP Work Plan);

- WRP partners should utilize the Project Tracker tool of EcoAtlas (ptrack. ecoatlas.org) and the WRP Marsh Adaptation Planning Tool (MAPT) (scwrp. databasin.org) to compile and share project information and data for all of the recommended monitoring components;

- The WRP should use standardized, state-of-the-science monitoring methodologies that enable comparison across projects and to regional and statewide monitoring;

- Monitoring protocols, designs, methods, and findings need independent scientific peer review;

- Regional management of monitoring data, with adequate data quality control and assurance, that allows data to be shared and publicly visualized; and

- Regular program review and revision to guide adaptation of the program in response to new scientific understanding, changes in technology, and changes in the WRP information needs. ●

FIDDLER CRAB AT NEWPORT BAY • PHOTO BY TRACIE HALL, COURTESY OF CREATIVE COMMONS

GOAL 2:

VERNAL POOL AT MONTGOMERY FIELD • PHOTO BY JOANNA GIKESON, COURTESY OF USFWS

Recovering Streams, Adjacent Habitats, and Other Non-Tidal Wetlands

Goal 2 of the *Regional Strategy 2018* is to *Preserve and restore streams, adjacent habitats, and other non-tidal wetland ecosystems to support healthy watersheds.* Although the primary focus of this document is the preservation, restoration, and management of coastal wetlands, the health of coastal wetlands is linked to material inputs and biological connections from the watersheds (Figure 28). Therefore, achieving the Objectives of Goal 1 requires restoration and management of the streams and non-tidal wetlands. The coastal watersheds and their streams, adjacent habitats and non-tidal wetlands (freshwater marshes, vernal pools, slope and seep wetlands, lakes, and non-tidal flats) contribute to watershed health in terms of providing habitat, improving water quality, and providing biogeochemical processes that support both aquatic and terrestrial wildlife.

Figure 28. Conceptual diagram demonstrating the hydrological connections and sediment sources between the watershed and coastal wetlands.

WETLAND LOSS AND CONVERSION

Historical changes to watersheds in Southern California have resulted from agricultural production, mineral extraction, water capture and diversion, urbanization, and infrastructure development. These impacts have resulted in changes to the magnitude, duration, and timing of freshwater flows, and the direct loss of wetland habitats. Particularly in large watersheds, natural stream flows were the first to be captured and diverted resulting in an overall loss of non-tidal wetlands. The urban runoff and wastewater discharges from those actions have also resulted in a redistribution of non-tidal wetlands. Watershed stressors have also disrupted physical and ecological processes and connections, and reduced the long-term resiliency of the watershed and the associated coastal estuaries. A majority of Southern California watersheds are considered to have moderate to high levels of vulnerability to climate change based on the California Integrated Assessment of Watershed Health (U.S. EPA 2013) because climate change is likely to change hydrographs, flow patterns, and habitats in streams and wetlands, resulting in flashier flows and generally drier and hotter conditions (promoting fire conditions).

APPROACH TO GOAL 2 OBJECTIVES

The Objectives for Goal 2 were developed by characterizing a range of watershed conditions based on state and regional monitoring data, limited historical data, and expert opinion. Whereas the tidal wetland Objectives (Goal 1) were based on an analysis of historical losses and projections of potential future losses due to sea level rise, comparable information was not available for the associated watersheds, non-tidal wetlands, and riparian areas. There has been no comprehensive analysis of historical non-tidal wetlands and riparian habitats completed for Southern California. Moreover, potential habitat changes due to climate change-induced alteration of rainfall patterns is much less predictable than sea-level rise, and regional projections have not been developed. Consequently, unlike Goal 1, the Objectives for Goal 2 were not based on a numeric comparison of the historical distribution of non-tidal wetlands to projections of potential changes from climate change, as this information is not as readily available. The Goal 2 Objectives were developed using an analysis of the range of conditions that currently exist, from most intact to most impacted habitats, available historical information, State and regional monitoring data, and expert input from the Science Advisory Panel.

In addition to the Key Concepts stated in the beginning of this document (page 8), the following additional Key Concepts apply to the Goal 2 Objectives:

- The Objectives are broadly inclusive of all aquatic resources in the coastal draining watersheds (headwaters to ocean).

- Development of the Objectives does not account for potential changes associated with climate change, such as changing temperature, runoff patterns, groundwater depths, fire frequencies etc., because these effects are much more difficult to predict and less certain than for coastal wetlands where sea-level rise is the predominant effect. Nevertheless, climate change is expected to impact the potential for non-tidal wetland restoration in the watersheds, and this issue should be explored further.

SANTA ANA RIVER • PHOTO BY DANIEL ORTH, COURTESY CREATIVE COMMONS

EXPLAINING NON-TIDAL WETLAND TERMS

Non-tidal wetlands • These wetlands include freshwater marshes, vernal pools, slope and seep wetlands, lakes, and non-tidal flats in the coastal watersheds.

Streams and adjacent habitats • The stream, its floodplain, and additional upland buffer habitat. Collectively, these habitats are sometimes referred to as the "stream corridor." This is measured in acres because it is being assessed through a wetland habitat lens.

Watershed • Land area that channels rainfall and snowmelt to creeks, streams, and rivers, and eventually to outflow points such as reservoirs, bays, and the ocean.

See Figure 29 below.

BELL CREEK AND SURROUNDING HABITAT • PHOTO COURTESY OF THE CALIFORNIA STATE COASTAL CONSERVANCY

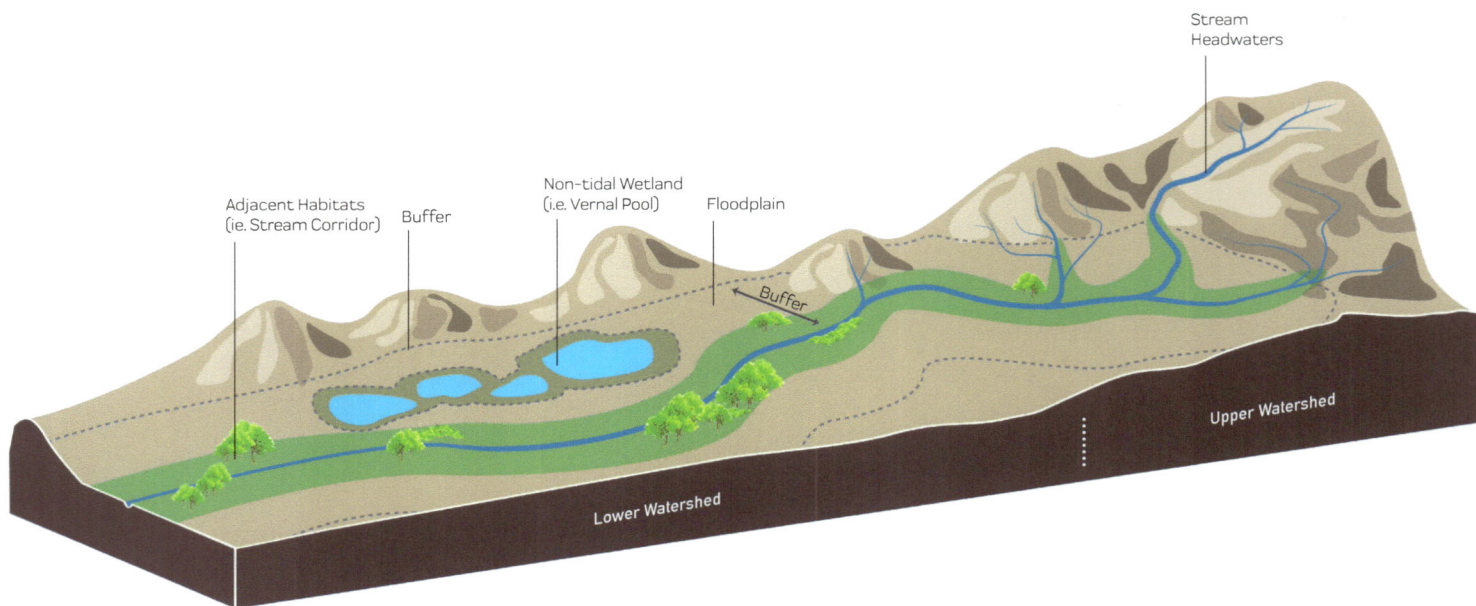

Figure 29. Conceptual diagram showing stream, adjacent habitats (riparian), buffer, and other non-tidal wetlands (vernal pools, lakes, etc.).

OBJECTIVES TO RECOVER STREAMS, ADJACENT HABITATS AND OTHER NON-TIDAL WETLANDS

The WRP has developed three Objectives, and a set of Management Strategies, to support the achievement of Goal 2 (Table 8). Achievement of the Goal 2 Objectives will help remedy some of the past and current impacts to wetlands and enhance the resiliency of Southern California's watersheds to future impacts. The Objectives and Management Strategies will provide measurable targets to evaluate our progress, guide the design of individual projects, and aid in project prioritization. The Goal 2 Objectives are applicable to the entire WRP region; no subregional Objectives are proposed. However, the regional Objectives can provide context for local decision making and project planning and design.

A detailed description of each Objective can be found below. For each Objective the same structure is used:

- Objective;

- Rationale for the Objective; and

- Recommended methodology to track the Objective.

Table 8. A summary of the Objectives that comprise Goal 2 of the Regional Strategy 2018.

Objective	Description
1. Streams, adjacent habitats, and other non-tidal wetland area	Maintain 160,618 acres (65,000 hectares) and restore 49,421 acres (16,766 hectaress) to achieve 210,039 acres (85,000 hectares) of non-tidal wetlands.
2. Habitat composition	A. Restore or maintain 189,036 acres (76,500 hectares) of streams and associated adjacent habitat. B. Restore or maintain 21,004 acres (8,500 hectares) other non-tidal wetlands (depressional, slope, etc.).
3. Connectivity	A. Ensure that there are no artificial physical barriers that obstruct water, sediment, and wildlife movement from watersheds to coastal wetlands. B. Remove 100% of the total and partial barriers to steelhead passage in the high priority watersheds identified in the Southern California Steelhead Recovery Plan (California Fish Passage Assessment Database (PAD)).

Protect and maintain 160,618 acres (65,000 ha) of existing non-tidal wetlands and restore an additional 49,421 acres (20,000 ha) to achieve a total of 210,039 acres (85,000 ha) of non-tidal wetlands.

Rationale: Maintaining an abundance (area) and diversity of wetland types will support aquatic dependent species (plants and animals) and associated ecosystem functions. Southern California has experienced substantial loss of wetlands and riparian systems. These losses have been associated with loss of function and adverse impacts to downstream coastal wetlands due to altered inputs of water, sediment, and organic matter, and the fragmentation of biological connections. Increasing overall wetland area will partially offset the historical losses. Most opportunities will be associated with expansion of riparian zones and floodplain wetlands.

The Objectives for protection and maintenance of non-tidal wetland areas were calculated using National Wetland Inventory (NWI) maps of current wetland extent, assuming that all existing wetland areas will be protected. The average density (area of wetlands per unit area of landscape) of non-tidal wetlands is 2.5%. We multiplied this current wetland density by the total coastal watershed land area of Southern California (approximately 6,380,000 acres [2,582,000 ha]) to arrive at an Objective to protect 160,618 acres (65,000 ha)of current wetlands. We used this method of multiplying wetland density by watershed area because there is no up-to-date comprehensive map of non-tidal wetland extent in Southern California.

The restoration targets were based on an examination of wetland distribution in a subset of relatively intact watersheds in the region and then extrapolated to the entire region. The

Figure 30. Historical habitats of the Santa Clara River and valley, early 1800s (Adapted from Beller et al. 2011).

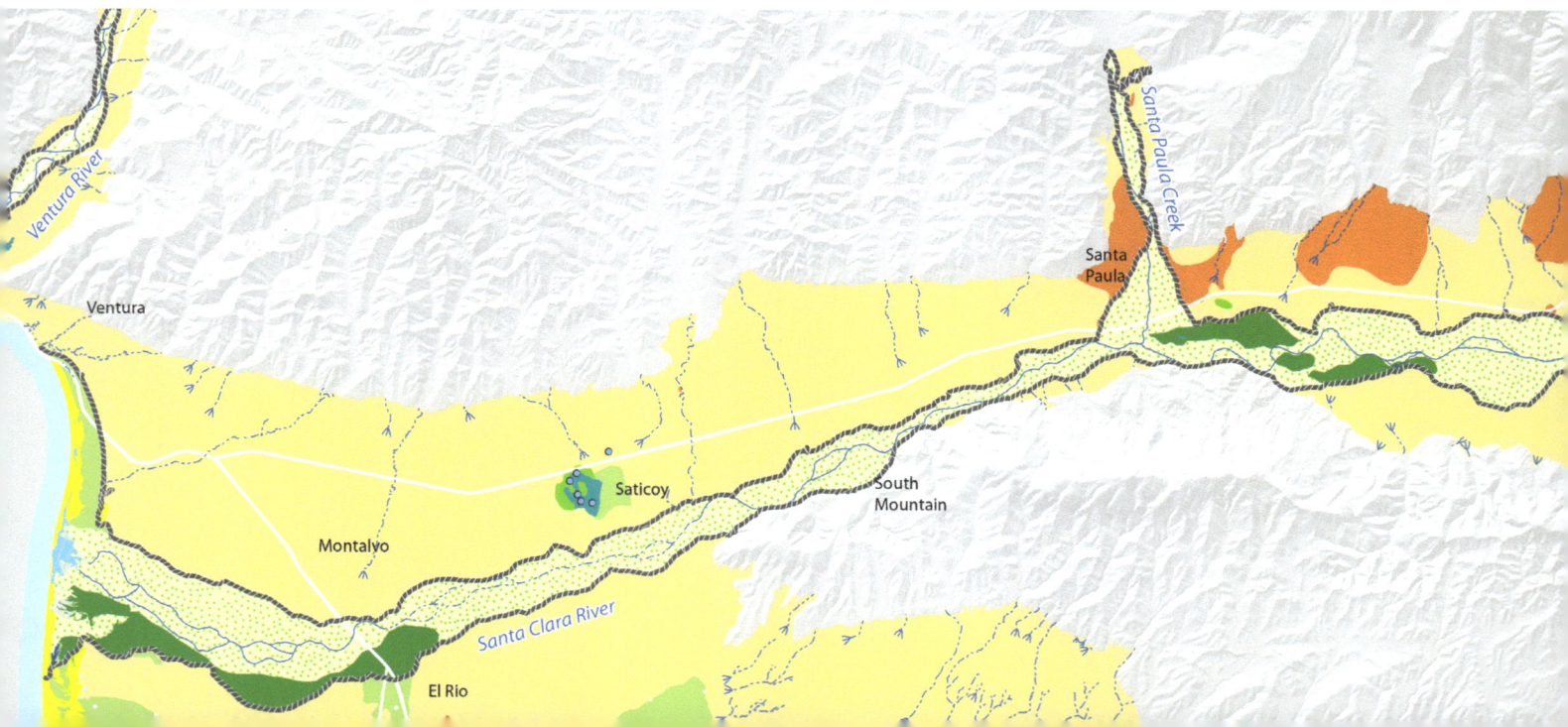

restoration targets were also guided by a limited number of historical ecology studies that demonstrate the pre-development landscapes of the San Gabriel River, Ballona, Ventura and Santa Clara River watersheds (Beller et al. 2011; Dark et al. 2011; Stein et al. 2007). See Figure 30 for an example along the Santa Clara River of the historical, non-tidal habitats. The average density of historical non-tidal wetlands, based on these studies, was 4.25% (range of 4% to 4.5%) and the estimated average historical coverage of non-tidal wetlands was 271,816 acres). By choosing an Objective for wetland density of 3.3%, which falls between the historical and the current wetland density, our Objective is to realize about 210,040 acres (85,000 ha) of non-tidal wetlands. In order to achieve this Objective we will need to restore approximately 49,421 acres (20,000 ha) of additional non-tidal wetlands.

Objective Tracking:

- Quantify the area of non-tidal wetlands that are protected or restored through Work Plan projects.

- Areal estimates should be reported by using the wetland types identified in the California Rapid Assessment Method (CRAM) (e.g., riverine, depressional, vernal pool, slope seep), which are also used in the California Aquatic Resource Inventory (CARI) (CWMW 2014).

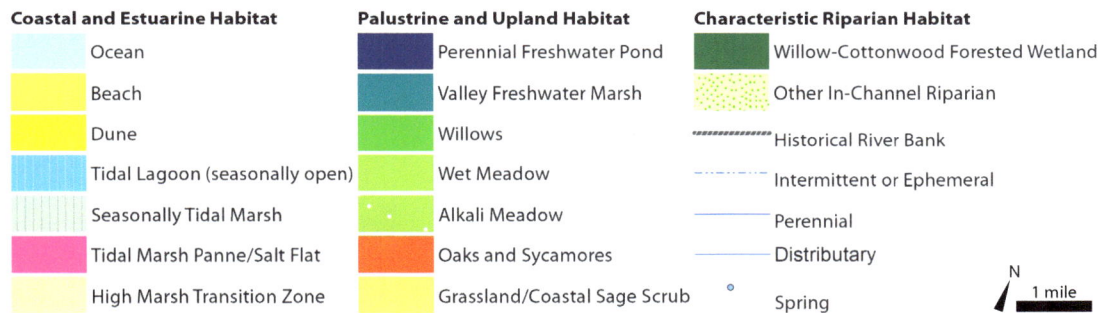

Coastal and Estuarine Habitat
- Ocean
- Beach
- Dune
- Tidal Lagoon (seasonally open)
- Seasonally Tidal Marsh
- Tidal Marsh Panne/Salt Flat
- High Marsh Transition Zone

Palustrine and Upland Habitat
- Perennial Freshwater Pond
- Valley Freshwater Marsh
- Willows
- Wet Meadow
- Alkali Meadow
- Oaks and Sycamores
- Grassland/Coastal Sage Scrub

Characteristic Riparian Habitat
- Willow-Cottonwood Forested Wetland
- Other In-Channel Riparian
- Historical River Bank
- Intermittent or Ephemeral
- Perennial
- Distributary
- Spring

1 mile

Protect or restore the non-tidal wetland composition to achieve 189,036 acres of wetlands associated with rivers and streams and 21,004 acres of other non-riverine wetlands.

Rationale: River-and stream-associated wetlands and non-riverine wetlands provide different landscape functions and may support different species or different life-stages of resident species. Both are important for overall watershed function. Restoration and management efforts should strive to maintain the appropriate distribution of riverine and non-riverine wetlands in the region.

Currently, 85% to 90% of non-tidal wetlands across all Southern California watersheds are riverine. The distribution of riverine versus non-riverine wetlands is consistent across the region, regardless of level of alteration in the watershed. Of the 210,040 acres (85,000 ha) of wetlands in Objective 1, 90% of riverine wetlands would result in 189,036 acres (76,500 ha) and 21,004 acres (8,500 ha) should be non-riverine.

Objective Tracking

- Quantify the area of non-tidal wetlands that are protected or restored through Work Plan projects (the same tracking measures as Objective 1)

- Areal estimates should be reported by CRAM/CARI wetland type (e.g., riverine, depressional, vernal pool, slope seep).

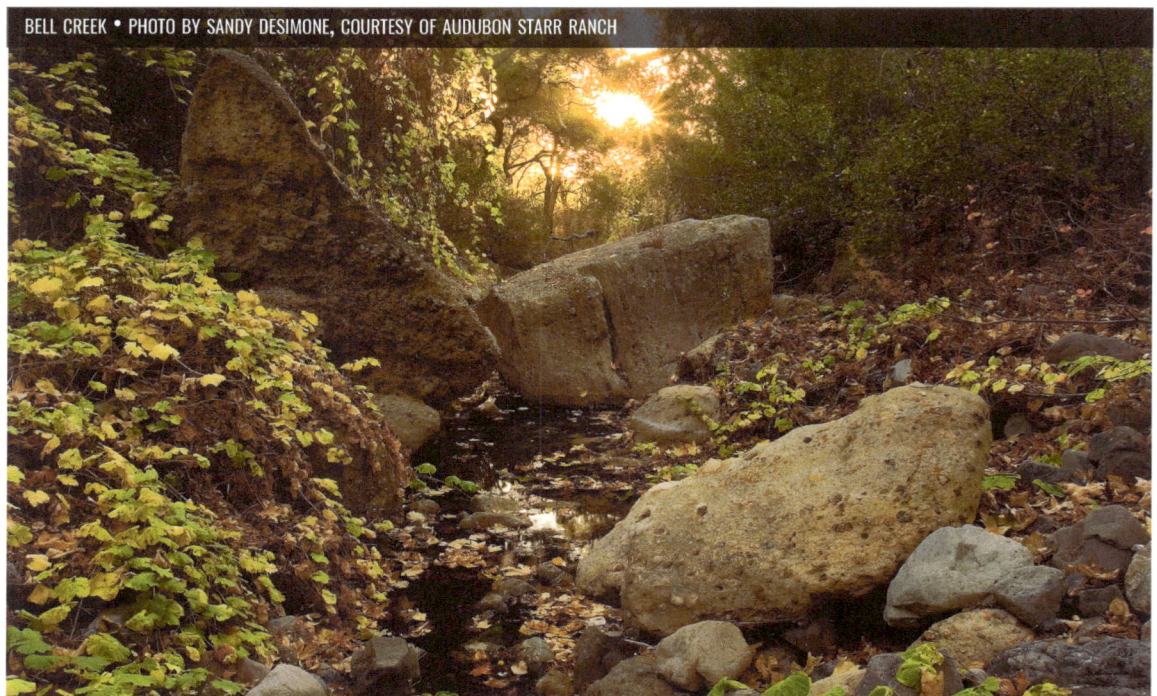

BELL CREEK • PHOTO BY SANDY DESIMONE, COURTESY OF AUDUBON STARR RANCH

A. Ensure that there are no artificial physical barriers that obstruct water, sediment, and wildlife movement from watersheds to coastal wetlands.

B. Remove or modify 100% of the total and partial barriers to steelhead passage in streams or rivers with high priority (Core 1-3) populations (Figures 31 and 32), as identified in the Southern California Steelhead Recovery Plan (PAD).

Rationale:

A. Reducing and eliminating barriers between coastal wetlands and lower watershed areas is consistent with *Regional Strategy 2018* Goal 1, Objective 6B: Restore freshwater and sediment flow characteristics from watersheds (volume, frequency, and timing) guided by appropriate reference systems, to support habitat abundance and distribution as indicated in Objectives 1–4. Protecting and restoring watershed connections helps support healthy coastal wetlands by maintaining important habitat linkages, migration routes and refugia for anadromous fish, migratory birds, and other coastal species. Maintaining watershed connectivity from the headwaters to the coast also protects appropriate water, sediment, and nutrient levels and periodicities.

B. The Objective was developed based on a goal of ultimately removing 100% of the known barriers to fish passage, which also serve to impede natural water and sediment movement. The Southern California Coast steelhead population is a federally-endangered distinct population segment (DPS). The recovery of this anadromous fish species depends on its ability to access freshwater spawning areas in coastal watersheds. Many spawning areas throughout Southern California are inaccessible to steelhead because of barriers to fish passage. Removing total and partial barriers to fish passage is an achievable Objective that will not only facilitate the recovery of steelhead and other fish, but will also improve the flow of water and sediment from the headwaters to the coast.

The PAD (https://map.dfg.ca.gov/bios/?al=ds69) estimates that there are over 830 partial or total barriers to fish passage in Southern California (Figures 31 and 32), with 189 barriers in core 1–3 streams, and many of these barriers also act to alter or obstruct flow patterns and sediment movement. The PAD serves as an up-to-date resource on mapped barriers and can be used to track progress toward the Objective of 100% removal of high priority fish-passage barriers in Southern California. See Goal 4 for research needs regarding the unassessed fish barriers identified in the California Passage Assessment Database.

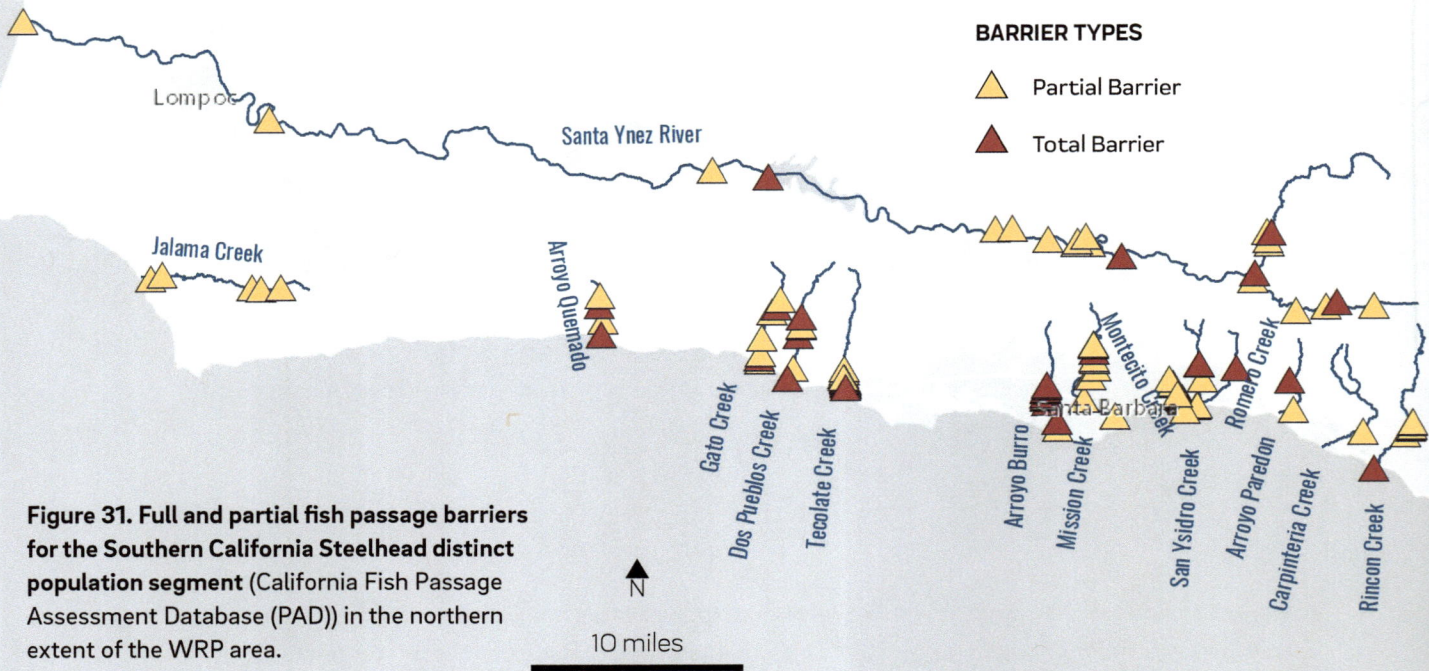

BARRIER TYPES

△ Partial Barrier

▲ Total Barrier

Figure 31. Full and partial fish passage barriers for the Southern California Steelhead distinct population segment (California Fish Passage Assessment Database (PAD)) in the northern extent of the WRP area.

N

10 miles

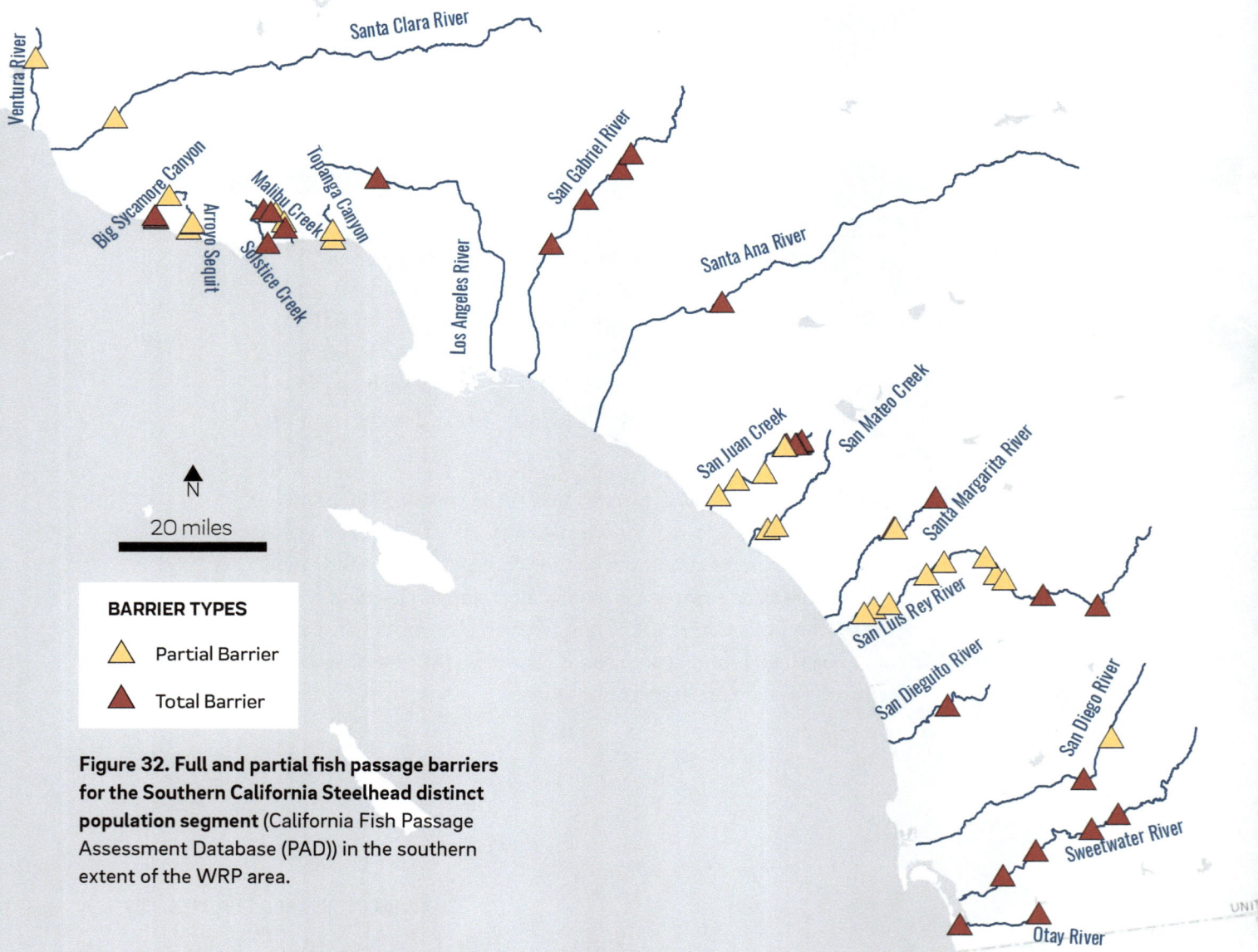

BARRIER TYPES

△ Partial Barrier

▲ Total Barrier

Figure 32. Full and partial fish passage barriers for the Southern California Steelhead distinct population segment (California Fish Passage Assessment Database (PAD)) in the southern extent of the WRP area.

N

20 miles

- Quantify the number of Work Plan projects that improve habitat, water or sediment connections between watersheds and coastal wetlands, and report how the project changed these connections.

- Quantify the number of Work Plan projects that involve removal or modification of fish passage barriers such as stream crossings, bridges, and culverts in order to improve steelhead habitats. Reporting should include a location map and a description of how passage was improved by the project.

- Work Plan projects involving removal of fish passage barriers should update the barrier information on the PAD website (https://map.dfg.ca.gov/bios/?al=ds69)

MANAGEMENT STRATEGIES

The following Management Strategies are intended to maintain watershed processes that promote health and resiliency of wetlands. Achieving the Objectives under Goals 1 and 2 will be helped by incorporating these Management Strategies into watershed management programs.

1. For all non-tidal areas restored, protected, and maintained, secure sufficient water resources to support healthy non-tidal wetlands. Non-tidal wetlands are dependent on a range of water sources (e.g., direct precipitation, groundwater, surface flows) and hydrographs to support a variety of vegetation communities and associated wildlife.

2. Maintain native habitat buffers adjacent to streams and wetlands.

- Buffers are defined as an area of native habitat beyond the riparian zone that protect the stream or wetland from the effects of adjacent land use. Buffers provide a transition zone between streams, adjacent habitats, non-tidal wetlands, and the adjacent uplands and have been shown to improve resiliency of streams and wetlands by reducing pollutant and excessive sediment loading from adjacent upland areas (Sweeney and Newbold 2014). Buffers can also provide area for seasonal expansion of wetlands, can allow for expansion during large, episodic storms, and can increase the wildlife function of streams and wetlands by providing habitat for various life stages of aquatic organisms.

- Appropriate buffer size will be related to the size of the stream or wetland relative to its surrounding landscape. For example, a small headwater stream in a forested landscape will need less buffer than a wetland in an urban setting. Buffers of approximately 30 m have been shown to substantially reduce water quality effects associated with runoff into the stream or wetland (Sweeney and Newbold 2014); however, up to 100 m may be necessary for trapping of fine sediments (Wenger 1999). Aquatic wildlife protection (primarily reptiles and amphibians) typically requires buffers of 100—300 m, but a buffer of up to 1,000 m may be required for some birds or mammals (Hruby 2013). It is important that buffers be managed along with the stream/wetland and adjacent riparian zone to minimize non-native invasive plant species.

3. Protect connections between critical coarse sediment areas and streams and maintain adequate sediment transport capacity through all mid- and low-order stream reaches so that coarse sediment can be delivered to the coast. Critical coarse sediment areas are still available in many watersheds. Protecting these areas can contribute to improved geomorphic condition.

4. Maintain natural flow patterns in 3rd order streams and higher (based on stream orders from Strahler 1952). Approximately 45% of streams are hydrologically altered to some degree in terms of their ability to support healthy benthic communities. Alteration is most pervasive in the lower portions of the watersheds, where the 3rd order streams occur.

5. Protect and manage streams to be in Channel Evolution Model (CEM) classes 1 or 5 (Hawley et al. 2012; Schumm et al. 1984). Class 1 is considered "unaltered." Class 5 streams have been previously altered, but have achieved a new equilibrium state based on the current hydrology. Both Classes 1 and 5 are considered relatively stable (unless hydrology is altered again) and both classes can be expected to support healthy biological communities. Approximately 75% of streams in the region experience moderate to high levels of incision or other geomorphic alteration.

6. Non-tidal habitat composition for a project should be determined by designing restoration actions that are appropriate for the current and future (i.e., projected) hydrology.

7. Protect open space in 0 and 1st order basins. Based on our GIS analysis, approximately 80% of 0 and 1st order sub-basins are open space in our most intact watersheds. Headwater basins are critical sources of sediment and organic matter that support biogeochemical functions lower in the watershed. Cumulatively, they also serve as important infiltration areas that support downstream hydrology. Restoration of healthy watersheds and resiliency of higher order streams depends on intact headwater basins.

8. Maintain natural non-perenniality in 1st through 3rd order streams. Up to 80% of stream reaches in Southern California watersheds are naturally non-perennial, but this has been altered by an increase in urban runoff, which creates perennial freshwater flow in many streams where it would not naturally occur.

9. Maintain groundwater levels close enough to the surface to maintain wetland and riparian vegetation communities in areas of existing shallow groundwater.

CONCLUSIONS

Maintaining healthy watershed structure and processes is imperative to ensuring long term resiliency of coastal wetlands and achieving the Objectives of Goal 1. Restoration and management of non-tidal wetlands will provide important aquatic species habitat and improve water quality for streams, adjacent habitats, other non-tidal wetlands, and coastal wetlands. The Goal 2 Objectives will help ensure persistence of an interconnected mosaic of freshwater wetlands in coastal watersheds and maintenance of critical functions necessary for the health of coastal wetlands. ●

BIOLOGISTS FROM MULTIPLE AGENCIES WORKING TO REMOVE NONNATIVE FISH FROM SANTA ANA SUCKER HABITAT • PHOTO COURTESY OF USFWS

GOAL 3:

The third Goal of the *Regional Strategy 2018* is, to *support education and compatible access related to coastal wetlands and watersheds.* The WRP has developed five Objectives and a set of related Management Strategies to support achievement of this Goal. The Objectives and Management Strategies provide targets to evaluate our progress and aid in project prioritization. Since implementation of the Objectives relies on project proponents to include education and access components in their wetland projects where appropriate and practicable, the objectives are not quantitative. It is up to the individual project proponents to set their own numeric targets for success. The Wetland Advisory Group and the Wetland Managers Group developed the Objectives for Goal 3 using best professional judgement gained through the collective experience of community groups.

Coastal Wetland Education
and Compatible Access

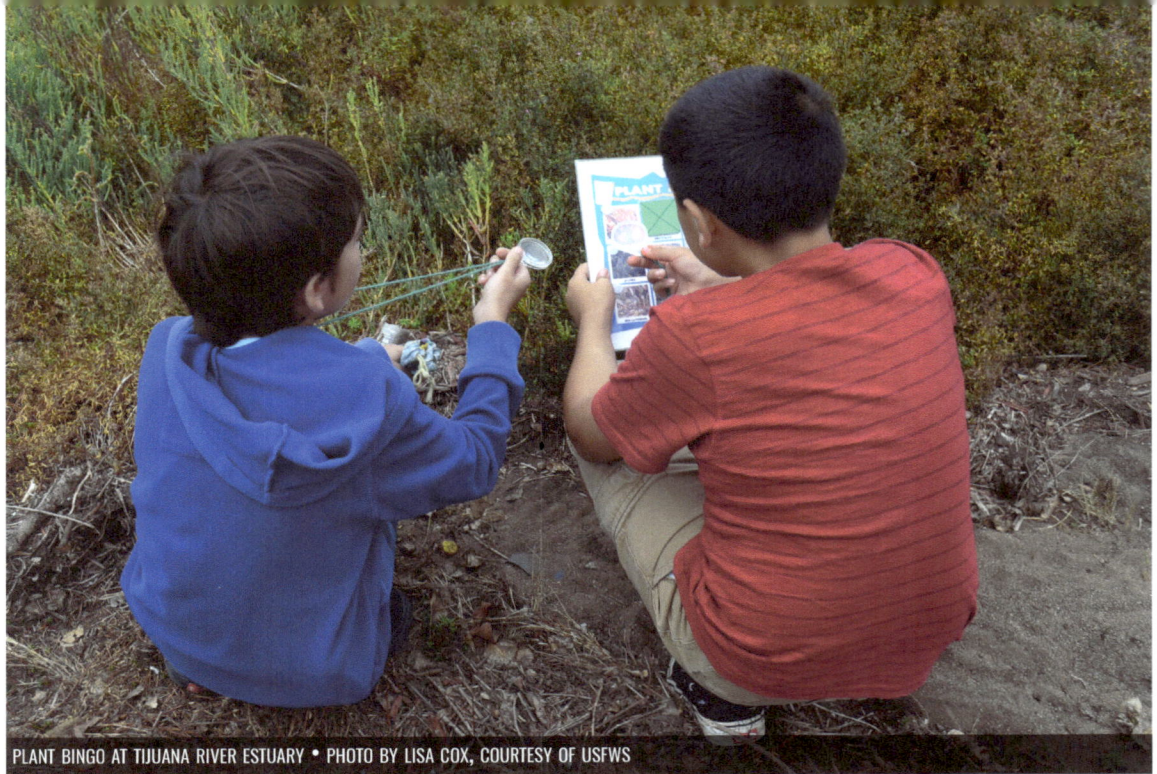
PLANT BINGO AT TIJUANA RIVER ESTUARY • PHOTO BY LISA COX, COURTESY OF USFWS

The Goal 3 Objectives were developed with the expectation that they will be accomplished through existing WRP initiatives. These initiatives include:

- WRP Work Plan projects;

- Community Wetlands Restoration Grants Program (CWRGP) projects;

- The WRP website; and

- Educational workshops, conferences and symposia.

The five Objectives address compatible public access, CWRGP projects, disadvantaged and underserved communities, interpretive programs and educational materials, web-based education, and conferences, workshops and symposia. The subject of a particular Objective may also be a subset of another Objective; for example, a CWRGP project may work in a disadvantaged community. The topic and specific language of each Objective are summarized in Table 9.

A more detailed description of the Objectives and the rationale supporting them is provided in the following pages. Additionally, while not setting measurable targets for each Objective, we provide a means of determining whether or not a specific Objective is being achieved.

Table 9. Summary of Goal 3 Objectives.

OBJECTIVE
1. Support community-based restoration projects.
2. Include compatible and equitable public access.
3. Integrate interpretive programs.
4. Promote development of educational materials and activities.
5. Disseminate wetlands science, research, and lessons learned.

Objective 1: Support Community-Based Restoration Projects

Rationale: Community-based projects build local capacity to plan and implement wetlands restoration, promote community involvement, and foster education about wetland ecosystems. Community-based projects are required to include strong educational and community involvement components.

The CWRGP helps to further the Goals of the *Regional Strategy 2018* in several ways and will specifically support achieving Objective 1. Each year, the CWRGP solicits proposals from nonprofit organizations, university departments, local government agencies and other eligible organizations. Proposals are reviewed by staff from WRP partner agencies. The CWRGP typically funds 10 to 12 projects per year. Over the past 16 years, the CWRGP has provided funding for more than 150 community-based projects.

Objective Tracking: In order to evaluate whether the WRP is achieving the CWRGP Objective, the WRP should track the following measures of success:

- Number of CWRGP projects completed each year;

- Number of participating volunteers and students;

- Acres and stream miles restored;

- Number/area/mass of invasive plants removed; and

- Number of native plants installed.

PREPARING EELGRASS FOR PLANTING • PHOTO COURTESY OF ORANGE COUNTY COASTKEEPER

Objective 2: Include Compatible and Equitable Public Access

Rationale: Our wetlands provide spaces of respite and inspiration within the highly urbanized landscape. Equitable access for all Californians to these places must also be balanced with preserving the sensitive habitats and species that epitomize them. Thoughtful and informed project design along with adequate resources for management can provide both public access and ecological preservation through considerations such as territory sizes, foraging patterns, and nesting seasons, and other annual patterns of species use. There is a critical balance between access and ensuring sensitive habitats and species are not impacted by access. Access may sometimes include physical access via trails, boardwalks, kayaks and other means. In other instances overlooks and viewing platforms may provide visual access without actual physical access.

Objective Tracking: In order to evaluate whether the WRP is achieving the compatible access Objective, the WRP should track the following measures of success:

- Number of projects integrating or improving access; and

- Number of projects decommissioning unintended and/or inappropriate access.

Objective 3: Integrate Interpretive Programs

Rationale: The WRP supports the development of programs that provide meaningful ways for people to visit and appreciate Southern California's wetlands. Through the establishment of interpretive programs that are integrated into restoration projects, our project partners can provide numerous opportunities for public participation that also help meet and support long-term care and maintenance needs. Interpretive programs range from citizen science monitoring programs and clean-up events to annual bird counts and tours. Through hands-on education and restoration experiences, children and adults alike are more likely to form a connection and commitment to preserving and restoring our coastal wetland resources.

Objective Tracking: In order to evaluate whether the WRP is achieving the interpretive programs Objective, the WRP should track the following measures of success:

- Number of WRP projects that integrate interpretive programs into project implementation or as a program following restoration.

Objective 4: Promote Development of Educational Materials and Activities

Rationale: In addition to the interpretive programs identified under Objective 3, educational materials and activities can help the public understand and develop connections to our wetland resources. These can range from informational flyers and species identification handouts or placards, to interpretive signage, to wildlife viewing scopes and audio tour programs. Projects that include the development and dissemination of such materials and activities help to develop a deeper understanding of wetlands among our Southern California communities.

Objective Tracking: In order to evaluate whether the WRP is achieving the educational materials and activities Objective, the WRP should track the following measures of success:

- Number of projects developing or disseminating educational materials and/or activities; and

- Numbers of people receiving educational materials and participating in activities.

Objective 5: Disseminate Wetlands Science, Research, and Lessons Learned

Rationale: The WRP has engaged in and led efforts to support the identification and funding of wetlands research needs in Southern California over the past 20 years. This research has included a wide body of historical ecology and the development of wetland monitoring protocols.

In addition, the WRP's many restoration partners have extensive knowledge and experience. The WRP continues to support the development of regional conferences and symposia to share wetlands research and lessons learned from developing and implementing restoration projects. The WRP collaborating agencies have also organized webinars around Southern California wetland topics. This role could be expanded in the future to disseminate information about the Regional Strategy, wetlands research, restoration issues and lessons learned from implementation of WRP supported projects.

The WRP website should serve as a source for public access to links to WRP-related research organizations and to see the latest research topics. By establishing links to organizations such as the Southern California Coastal Water Research Project, San Francisco Estuary Institute-Aquatic Science Center, and the Tijuana Estuary National Estuarine Research Reserve, the WRP can help our partners and the public access the latest research and reports.

Objective Tracking: In order to evaluate whether the WRP is achieving the science and research dissemination Objective, the WRP should track the following measures of success:

- The extent to which the list of research organizations linked from the WRP website is comprehensive;

- Number of research publications and projects promoted on the WRP website; and

- Number of WRP project proponents and partner agencies planning and organizing, participating in, or presenting at wetland related conferences and symposia.

OYSTER BED CONSTRUCTION • PHOTO COURTESY OF ORANGE COUNTY COASTKEEPER

ACTIVITIES

The Goal 3 Objectives are focused on initiatives that currently exist within the WRP. There are also a number of educational and outreach initiatives beyond those that have been the historical focus of the WRP. While these education and compatible access initiatives fall outside the scope of WRP partner agency efforts, they are worth noting here as activities that local partners could develop and implement. These activities include:

PARTNERING:

Promote partnering between more experienced and less experienced project proponents, allowing less experienced partners to benefit from the lessons learned of organizations with greater capacity.

EDUCATION STANDARDS:

Utilize educational materials that interface with Next Generation Science Standards (NGSS) for educational initiatives that are focused on students and youth.

EDUCATIONAL TECHNOLOGY:

Develop smartphone apps that can be used at wetland sites as self-guided interpretive tools or as a means to facilitate citizen science.

EVENTS:

Hold community events at wetland sites as a means to engage and educate the general public. Possible events include 5k and 10k walks and fun runs, lecture series, guided bird watching, farmer's markets, painting and photography workshops.

FUNDING:

Seek funding to support the Goal 3 Objectives. Funding for the WRP's Work Plan and CWRGP projects has historically come from the WRP partner agencies' grant programs. These funding sources typically do not fund educational and outreach initiatives, unless as a secondary component of a restoration project. The WRP will continue to look to local partners to increasingly and innovatively engage in educating the public on Southern California wetland issues. Through engaging the public and promoting wetland visitation our partners can help direct cultural and behavioral changes around our wetlands.

GOAL 4:

In pursuit of Goal 4 to *Advance the science of wetland restoration and management in Southern California,* the WRP has identified several regional knowledge gaps that, if filled, would improve our collective ability to manage the processes governing Southern California's wetlands.

The *Regional Strategy 2018* should be periodically updated to reflect advances in wetland and climate change science, new policies and community needs, and shifting opportunities and constraints in the landscape. We recommend completing the next *Regional Strategy* update by 2030 as that will provide the scientific community time to address many of the identified research needs. Also, 2030 is one of the first milestones in the Goal 1 timeline (Figure 26) where much of the land required to meet our Goal 1 Objectives should be in public ownership or protected. By updating the *Regional Strategy 2018* at this first milestone, the WRP will be able to assess progress in achieving its Objectives, incorporate new science, and adjust Objectives to meet new needs if necessary. The research questions and *Regional Strategy 2018* needs outlined in this chapter should be addressed within the next 10 years in order to provide the information required to update the *Regional Strategy* again by 2030.

Advance the Science
of Wetland Restoration and Management

Although there are many knowledge gaps about climate change adaptation and wetland restoration science, the proposed list of research needs was directly derived from the development of the *Regional Strategy 2018*, and relates to research that is necessary to implement or track progress on the *Regional Strategy 2018* Objectives. During the *Regional Strategy 2018* development, the WRP identified the following five general research categories:

- Ecological Zones of Unique Southern California Wetlands

- Ecological Zones of Habitats Adjacent to Vegetated Coastal Wetlands

- Impacts of Hardened Coastal Infrastructure

- Future Wetland Conditions

- Wetland Restoration Best Practices and Procedures

The following research questions and priorities (i.e., the Goal 4 Objectives) were identified following the analyses performed for Goals 1 and 2. The Goal 4 Objectives are not quantitative, however the WRP will be tracking progress to answering the Objectives as those results will guide the 2030 update of the Regional Strategy 2018.

SURVEYING FOR SALT MARSH BIRD'S BEAK • PHOTO BY JOANNA GILKESON, COURTESY OF USFWS

Rationale: The WRP only provided a management strategy under Goal 1 for salt flat habitats because the information on contemporary and historical extent, distribution, and ecosystem functions of salt flats that would be needed to develop a quantitative Objective does not currently exist. Furthermore, the salt flat habitats that exist today are largely within disturbed systems and the driving processes in developing and maintaining the salt flats remain unknown (Appendix 7). While studies in other regions have demonstrated the use of salt flats by resident and migratory birds, invertebrates, and traveling mammals, it remains unknown whether fish species utilize these habitats during their seasonal and/or annual wet phases. It is likely that listed species, such as the Tidewater Goby and Southern California Steelhead, do seek refuge and forage in salt flats during the wet season, but data on this issue are not currently available. Future studies would enhance our understanding of the contemporary distribution of salt flats in the region and change over time, help refine the salt flat typologies, elucidate the ecological functions and services provided by these features, and increase our understanding of the formative processes and requisite conditions for their persistence. The specific research questions are:

- What is the existing and historical extent and distribution of salt flats?

- What is the change in salt flat extent and distribution over time?

- What are/were the ecosystem functions and services of salt flats during different seasonal and/or annual dry and wet phases?

- Do salt flat functions and services differ by wetland archetype?

- What are the landscape positions and formative processes (e.g. climate, tidal and freshwater influence, sediment dynamics, and elevation) that develop and maintain salt flats, both artificial and natural?

- Do listed fish species, such as the Tidewater Goby and Southern California Steelhead, use flooded salt flats as habitat?

- Do other resident brackish fish communities use flooded salt flats as habitat?

- How do bird communities make use of salt flat habitats throughout the year?

- What are the best restoration practices to restore salt flat habitats?

Rationale: Intermittently-open estuaries (IOE), tidal wetlands that close periodically to the ocean via a sandbar or beach berm that builds seasonally at the tidal inlet, are defining features of the Southern California coastal landscape. However, very few estuaries in Southern California still exhibit natural cycles of intermittency and many are managed as permanently open systems. With additional data, there could be opportunities to rethink the management of intermittently-open estuaries and the ecological trade-offs that go along with management decisions. In addition, rising sea levels will drive the need to consider management practices under new hydrological conditions.

The research questions below were developed from a workshop organized by the Tijuana River National Estuarine Research Reserve and were identified through the sea-level rise (SLR) modeling work presented in Appendices 3 and 6.

- What is the appropriate way to characterize "natural" inlet dynamics (e.g., closure periodicities and durations by archetype), and how can this be used to develop management approaches?

- How will the probability, timing, frequency and duration of inlet closures be affected by climate change, including SLR and its interaction with other forcing factors?

- How will changes in inlet closure timing, frequency, and duration affect key habitats (e.g., salt marsh), species (birds, steelhead, and tidewater gobies), and ecosystem processes/ services (e.g., carbon sequestration)?

- Can habitats, species, and services associated with closing systems be supported in open systems, and vice-versa?

- What is the efficacy in using living shoreline approaches to providing tidal inlet resilience in the face of sea-level rise?

- How can extreme events, such as king tides and El Niño, be used as a potential preview of climate impacts on inlet dynamics?

- What are the tradeoffs associated with passive versus active mouth management, and which systems might be most appropriate for more passive approaches?

- Which IOEs in Southern California would perform best with passive mouth management?

- Develop a model or tool to help local jurisdictions weigh the benefits and constraints for inlet management in terms of development/infrastructure vulnerability to flooding (industrial, commercial, residential), natural resource issues (e.g., steelhead, tidewater gobies), water quality issues, and historical ecology.

Rationale: The WRP did not develop quantitative Objectives for wetland restoration on the Channel Islands during the development of the *Regional Strategy 2018*. In the interim, the WRP developed a subcommittee of Channel Island wetland practitioners to understand the importance and state-of-the-science for these wetland systems (Appendix 12). The Channel Island subcommittee identified the following research questions:

- What is the historical extent of wetland resources on the Channel Islands?

- What are the wetland resources on Santa Cruz and Santa Rosa Islands?

- Do the WRP archetype classifications apply to Channel Islands wetlands?

- What are the potential wetland habitat changes in the face of seal-level rise on the Channel Islands?

SANTA CRUZ ISLAND • PHOTO BY DAVID FULMER, COURTESY OF CREATIVE COMMONS

Objective 4: Refine Objectives for Wetland–Upland Transition Zones

Rationale: *The Regional Strategy 2018* provides wetland-upland transition zone (hereafter "transition zone") Objectives under Goal 1 to promote transition zone protection and expansion. However, additional mapping and research is needed in order to provide a quantitative Objective for the amount of acquisition and restoration needed to restore transition zones (Appendix 9). The following research questions will help refine these Objectives:

- What is the extent of transition zone available above the marsh migration zone and what are the opportunities to create additional transition zone?

- What is the best approach for creating/restoring transition zone habitats?

- What is the relationship between transition zone topography and colonization by marsh vegetation?

- What are the physical and ecological characteristics of the transition zone that make it good habitat?

- What are the habitat functions currently provided by transition zones and how might those functions change as the transition zone becomes future wetlands?

Objective 5: Develop Quantitative Objectives for Shallow Subtidal Areas

Rationale: The WRP recognizes the importance of every habitat along an estuarine elevation gradient. However, the WRP could only provide a management strategy for shallow subtidal areas including soft bottoms, submerged aquatic vegetation, and reef/hardbottoms under Goal 1 due to a lack of expertise and available research (Appendix 8). For instance, maps of submerged aquatic vegetation (SAV) beds attributed by the National Wetlands Inventory are generally not useful because of the limitations of aerial photography.

Additionally, there are important research gaps in the ecosystem services valuation for shallow subtidal habitats. Current legislation being considered in California (e.g. SB 1363) explicitly calls for (1) Developing demonstration projects to research how important environmental and ecological factors interact across space and time to influence how geographically dispersed eelgrass beds function for carbon dioxide removal and hypoxia reduction, (2) Generating an inventory of locations where conservation or restoration of aquatic habitats, including eelgrass, can be successfully applied to mitigate ocean acidification and hypoxia, and (3) Incorporating consideration of carbon dioxide removal for eelgrass restoration projects during the habitat restoration planning process in order to fully account for the benefits of long-term carbon storage of habitat restoration in addition to the habitat value. This makes the pursuit of these research Objectives timely and more urgent.

Research priorities identified by the WRP include:

- What are the current distributions of shallow subtidal habitats (e.g., soft bottoms, submerged aquatic vegetation, and hard bottom/reef habitats)?

- What is the estimate of potential SAV habitat in Southern California (Bernstein et al. 2011)?

- What is the nursery role of various subtidal habitat types for juvenile fish species?

- What are the best management practices for the conservation of existing SAV habitats?

- What are the quantified ecosystem services for overlooked subtidal habitats such as soft bottom subtidal habitats?

  * How do important environmental and ecological factors interact across space and time to influence how geographically dispersed eelgrass beds function for carbon dioxide removal and hypoxia reduction?

  * Where have conservation and restoration of aquatic habitats, including eelgrass, been successfully applied to mitigate ocean acidification and hypoxia?

- What are the species, number, and distribution of subtidal invasive species?

- What are the best eradication practices for subtidal invasive species?

- What are the physical and biological conditions that support eelgrass and native oyster habitat?

- How will subtidal habitats change in distribution and function as sea levels rise?

- What are the ecological services provided by eelgrass ecotones like unvegetated shallow subtidal areas?

NATIVE OLYMPIA OYSTER • PHOTO BY VIUDEEPBAY
COURTESY OF CREATIVE COMMONS

Rationale: The quantitative Objectives for non-tidal wetlands developed for Goal 2 are a good first step toward a regional picture for the restoration and protection of the region's watersheds. However, the restoration Objectives for Goal 2 were not developed with the same level of information and rigorous analysis that was used for the quantitative Objectives in Goal 1. Through the development of restoration Objectives for Goal 2 the following data gaps were identified:

- What were the historical conditions for non-tidal wetlands?

- Where is the current distribution of non-tidal wetlands?

- Develop quantifiable habitat objectives for each type of non-tidal wetland (e.g., streams, freshwater marshes, vernal pools).

- What are the driving sediment dynamics for non-tidal wetlands?

- What will be the regional effects of climate change on non-tidal wetlands (e.g., changes in precipitation and freshwater runoff)?

- Which fish passage barriers affect sediment and water movement?

- Which barriers identified as "not assessed," "remediated, fish response unconfirmed," and "unknown passage status" in the California Passage Assessment Database serve as steelhead passage barriers?

- How do different buffer widths affect function and resiliency of non-tidal wetlands?

- How do changes in the condition of non-tidal wetlands affect the condition of downstream tidal wetlands?

SANTA CLARA RIVER • PHOTO COURTESY OF U.S. ARMY CORPS OF ENGINEERS

IMPACTS OF HARDENED COASTAL INFRASTRUCTURE
Objective 7: Analyze the Ecological and Physical Effects of Removing Coastal Infrastructure

Rationale: Even though a wealth of research has demonstrated the detrimental effects of hard infrastructure on coastal habitats and associated biological communities, little to no research has been conducted on the benefits of removing coastal infrastructure already in place. Southern California contains the most coastal hardening of any region in the nation (Gittman et al. 2015) and although many structures are necessary and provide essential physical protection, some structures could be replaced with natural or nature-based infrastructure. The following questions will help identify which coastal structures can and should be removed to help improve the coastal resilience and wetland functions of the region:

- What are the potential ecological benefits and impacts of removing coastal jetties and other hard infrastructure on wetland habitats?

- What are the feasibility, benefits and consequences of removing jetties and other infrastructure on erosion, flooding, sedimentation, and water quality?

- What factors, such as setting or habitat, are most important in determining the relative benefit and risk of removing hard infrastructure?

- What are the relative benefits and protection opportunities of soft armoring approaches (e.g. living shorelines) versus hard infrastructure?

- What are the conditions that support the use of living shorelines approaches in place of hard structures?

FUTURE WETLAND CONDITIONS
Objective 8: Refine the Vulnerabilities of Southern California Estuaries to Climate Change Impacts

Rationale: The WRP developed a numerical model to determine the magnitude of wetland habitat change due to the impacts of sea-level rise (Appendix 3). This provided a method to analyze all 90 wetland systems using the same level of data and analysis. During data collection for the model inputs, several data gaps were discovered. If the following data gaps were addressed, the model could be refined and re-run to better understand the potential impacts of sea-level rise to Southern California's coastal wetlands:

- What are the differences in sediment accretion rates, sediment sources and sediment processes between wetland types?

- What are the different elevations within various zones and habitats within coastal wetlands and how do these vary over time both within and between wetlands?

- How will the probability of marsh-dependent species (including threatened and endangered species) to complete various life history needs be affected by expected climate change factors, including changes in sea level, precipitation, and temperature?

Rationale: The Regional Strategy 2018 provides Objectives under Goal 1 to promote and facilitate marsh migration with sea-level rise. However, additional mapping and research is needed in order to provide a quantitative Objective for the amount of acquisition and restoration needed to restore and protect marsh migration areas. The following research questions will help refine these Objectives:

- What is the extent of available marsh migration areas and where are the opportunities to create additional marsh migration areas?

- What is the best approach for creating/restoring marsh migration areas?

- At what rate will sea levels be rising too fast for marsh vegetation to establish in the transition zones (i.e., outpace marsh migration)?

BOLSA CHICA • PHOTO BY SERGEI GUSSEV, COURTESY OF CREATIVE COMMONS

Objective 10: Identify the Best Restoration Practices for Southern California Coastal Wetlands and Non-Tidal Wetlands

Rationale: The WRP has been involved in wetland acquisition, enhancement, restoration, and education in Southern California for more than 20 years, and has built a wealth of experience and knowledge on wetland restoration practices over this time. Recently, Fejtek et al. (2014) developed a suite of best practices for coastal wetland restoration, which should be consulted throughout restoration planning processes. This manual of best practices did not include non-tidal wetlands, but a more recent guide has been developed to summarize best practices for steelhead restoration (Fejtek 2017). In addition, the WRP's Guiding Principles (page 10) memorialize some of the wetland restoration knowledge from the past 20 years. The following research questions will help the WRP continue to build upon lessons learned from the past, develop more specific project guidance, and remain at the forefront of restoration methods for the future:

DEVEREAUX SLOUGH, NORTHSHORE MARGIN ENHANCEMENT • PHOTO COURTESY OF SANTA BARBARA CO. AUDUBON SOCIETY

- What are the best methods for restoring native habitat diversity in tidal wetlands across different archetypes and settings?

- What are the best methods for restoring sediment and freshwater flow in coastal wetlands across different archetypes and settings?

- What are the best methods for restoring hydrology and habitat connectivity of coastal wetlands across different archetypes and settings?

- What are the characteristics of past successful and unsuccessful tidal and non-tidal wetland restoration projects across various scales and wetland types?

- How will institutional memory of restoration best practices be accumulated and shared through the WRP?

WETLAND RESTORATION BEST PRACTICES
Objective 11: Refine the Monitoring and Assessment Program to Track Progress Toward our Objectives

Rationale: Past studies of wetland mitigation and restoration have demonstrated that meeting permit or grant requirements does not equate with ecological success (Ambrose et al. 2006). One reason for this failure is that "monitoring" required as conditions of permits or grants focus on accounting of actions performed versus assessing actual ecological effects. Moreover, obtaining funding for long-term monitoring is difficult. Maximizing the likelihood of ecological success requires performance measures focused on process and function and long-term monitoring to ensure that these measures are achieved. Ongoing monitoring also facilitates adaptive management and implementation of remedial measures as necessary to ensure long-term ecological success and resiliency. The current strategy for tracking achievement of the Objectives for Goals 1 and 2 focuses on the contribution of projects on the WRP Work Plan. However, in order to successfully assess wetland health across the entire region, the WRP recommends developing a regional wetland monitoring program as described on page 77.

WETLAND RESTORATION BEST PRACTICES
Objective 12: Develop a Standing Science Advisory Panel for the WRP Region

Rationale: The Science Advisory Panel (SAP) that guided the development of the *Regional Strategy 2018* consisted of 11 scientists with a range of expertise in wetland restoration in Southern California and throughout the state. Due to a lack of adequate funding, the SAP will disband following completion of the *Regional Strategy 2018*. According to Fejtek et al. (2014), having a standing science advisory panel for the region would provide wetland practitioners with a reliable source of scientific guidance on the design, implementation, and management of wetland restoration projects. A standing SAP could serve the Southern California wetland restoration community in the following ways:

- Provide guidance to project proponents on how to utilize the *Regional Strategy 2018* report, Objectives, and tools in their project development;

- Reduce the burden on project proponents to develop their own technical advisory groups by serving as a regional technical advisory group for specific wetland restoration projects;

- Keep track of best restoration practices over time and across the region; and

- Exchange and disseminate scientific and technical information on restoration practices.

Glossary of Terms
for the *WRP Regional Strategy 2018*

Archetype – Coastal wetlands are complex and highly variable. Archetypes are representations of a group or class of objects (in this case coastal wetlands) of similar form and structure. Grouping wetlands into archetypes is useful because they provide a general conceptual model that can be used to explain how a specific group of wetlands function and how they may respond to external pressures or drivers. In this way, they help simplify analysis and communication, and provide a mechanism to generalize or extrapolate knowledge about a given system to similar systems (i.e., archetypes).

Regional Strategy 2018 – The WRP analyzed 58 variables related to physical conditions/drivers such as catchment properties, wetland dimensions, size and slope ratios, proportion subtidal versus intertidal, inlet dimensions and condition, and wetland volume/capacity to conduct a cluster analysis to group similar coastal wetlands. The cluster analysis was then refined by the Science Advisory Panel, using best-professional judgement, resulting in the following archetypes:

- Small creek
- Small lagoon
- Intermediate estuary
- Large lagoon
- Large river valley estuary
- Fragmented river valley estuary
- Open bay/harbor

Ambient Assessment – The characterization of regional (or statewide) conditions. Ambient assessments provide information on the extent, distribution, and condition of aquatic resources across a defined geography.

Anadromous Fish – A fish born in freshwater, but spends most of its life in the sea, and then returns to freshwater to spawn.

Basin Order – The area that drains to a point on the landscape where an incipient channel forms.

Bar-Built Estuary – See Intermittently Open Estuary below.

Benthic Communities – groups of organisms that live in and on the bottom of a water body floor.

Buffer – Areas adjacent to water resources to provide protection. Non-tidal wetland buffers provide a transition zone between streams/wetlands and the adjacent uplands and have been shown to improve resiliency of streams and wetlands by reducing pollutant and excessive sediment loading from adjacent upland areas (Sweeney and Newbold 2014). Buffers can also provide area for seasonal expansion of wetlands, can allow for expansion during large, episodic storms, and can increase the wildlife function of streams and wetlands by providing habitat for various life stages of aquatic organisms.

Channel Evolution Model – Idealized depictions of changes in the physical form of a stream change in response to disturbances. CEMs include a depiction of different states of a channel along a gradient of disturbance.

Conceptual Model – Narrative and graphics that articulate the physical and biological processes related to coastal wetland systems and their associated habitats. They describe our present understanding of processes and linkages, help anticipate and evaluate physical and biological responses to given scenarios, and identify unknowns and uncertainties of the system.

Coastal Wetlands – These are the coastal tidal wetland ecosystems that include shallow subtidal channels, vegetated marsh, unvegetated flats, and adjacent upland transitional areas. Coastal wetlands can also be part of open bays and harbors, but do not encompass the deeper subtidal areas.

Coherence – Connecting or reconnecting the natural flows of water and sediment within a coastal wetland system. In many cases these have been disrupted by the presence of berms and culverts.

Community Wetland Restoration Grant Program – The Community Wetland Restoration Grant Program (CWRGP) is a program of the California State Coastal Conservancy that provides funding annually for community-based restoration projects in coastal wetlands and watersheds in the Southern California region.

County Task Forces – There are five County Task Forces of the WRP. Each task force is co-chaired by a County Supervisor and an environmental leader. The Task Forces provide a county-wide forum for public, private, and non-profit wetlands and watershed stakeholders. Participants work collaboratively to identify critical wetland resources, help implement feasible projects, mobilize support for funding, channel community concerns to the WRP member agencies, incorporate wetlands protection and recovery more fully into local government processes, and promote wetlands education and information-gathering. The County Task Forces meet on an as-needed basis now that the Wetland Advisory Group (see below) has been developed.

Directors Group (DiG) – The coordinating body of the Southern California Wetlands Recovery Project (WRP). The DiG is comprised of the top officials from the 18 state and federal agency partners of the WRP and is chaired by the Secretary of the State Resources Agency.

Ecosystem Function –

- The interactions between organisms and the physical environment (Biology Online 2018).
- The biological, geochemical, and physical processes and components that take place or occur within an ecosystem (SEQ n.d.).
- The structural components of an ecosystem (e.g. vegetation, water, soil, atmosphere and biota) and how they interact with each other, within ecosystems and across ecosystems. Sometimes, ecosystem functions are called ecological processes (SEQ n.d.).

Ecosystem Service –
- Inextricably linked to ecosystem function; you cannot have an ecosystem service without the function.
- The conditions and processes through which natural ecosystems, and the species that make them up, sustain and fulfill human life (Daily et al. 1997).
- The benefits people obtain from ecosystems. These include provisioning services such as food and water; regulating services such as regulation of floods, drought, land degradation, and disease; supporting services such as soil formation and nutrient cycling; and cultural services such as recreational, spiritual, religious, and other nonmaterial benefits (Millennium Ecosystem Assessment 2005).
- Wetlands deliver a wide range of ecosystem services that contribute to human well-being, such as fish and fiber, water supply, water purification, climate regulation, flood regulation, coastal protection, recreational opportunities, and tourism (Millennium Ecosystem Assessment 2005).

Goals – Represents the overarching ambitions of the WRP.

Guiding Principles – Developed by the WMG, WAG, and SAP to assist in the development of the Goals.

Habitat Diversity – The variety of habitats (i.e., subtidal, unvegetated flat, vegetated marsh, & transition zone) in a coastal wetland system (see Coastal Wetland above).

Hypsometry – Measurement of land elevation relative to sea level (i.e., the tide). A hypsometric curve displays the elevation on the vertical axis and area below the corresponding elevation on the horizontal axis.

Incision – Deepening and/or widening of a stream channel in response to changes in flow and/or sediment yield.

Intermittently Open Estuary – An estuary with dynamic connections with the ocean that vary seasonally reflecting annual patterns of precipitation and river discharge as well as multi-year patterns of wet and dry years (see pages 51-53).

Marsh Adaptation Planning Tool (MAPT) – The site-specific online tool that the WRP will utilize to evaluate and prioritize restoration projects and that potential grantees will use to develop projects. The MAPT will help WRP partners translate the regional Goals down to site-specific recommendations. The MAPT is available at scwrp.databasin.org.

Native Habitat Buffer – Native vegetated area near a certain habitat that helps to protect the habitat from the impact of adjacent land-uses.

Natural – An ecosystem with the least amount of human disturbance compared to similar ecosystems in the region.

Non-tidal Wetlands – These wetlands include streams, riparian zones, freshwater marshes, vernal pools, slope and seep wetlands, lakes, and non-tidal flats in the watersheds that drain to the Southern California Bight.

Objective – These are quantifiable, spatially explicit, and time-bound measures, based on science, used to assess progress toward meeting the WRP Goals.

Outcomes – Changes expected as a result of the Regional Strategy 2018 implementation. These can be short-term, mid-term, or long-term.

Patch – The patch scale represents different habitat types such as vegetated marsh and unvegetated flats within a tidal wetland. It does not represent specific habitat zones such as low/mid/high marsh or specifics of plant species such as pickelweed-dominated marsh.

Preserve – Maintain an ecosystem to function in its natural state.

Reef/Hardbottom – Hard substrate in shallow subtidal areas composed of exposed bedrock or created through depositional cementation of sediment and includes corals and flat bedrock.

Reference Condition – The condition used as a basis of comparison to judge the performance of a restoration site. Depending on the project, the reference condition for a given variable can be anything from the pre-European condition to the best attainable condition given current constraints of the landscape. Defining the reference condition is a management decision that is made for each project.

Reference System – A site that reflects the least altered wetland in the landscape, and often the site used to compare reference conditions for project-specific monitoring (see Reference Condition).

Residence Time – Amount of water in an estuary divided by either the rate of addition of water to the estuary or the rate of loss from it.

Resilience – According to Walker et al. (2004), ecological resilience is "the capacity of a system to absorb disturbance and reorganize while undergoing change so as to still retain essentially the same function, structure, identity, and feedbacks." The Walker et al. definition was used for the RSU in that systems can adapt and change following disturbance (chronic or intermittent) as long as the primary ecosystem function and structure remains intact.

Restoration – Ecological restoration is the process of assisting the recovery of an ecosystem that has been degraded, damaged, or destroyed (SER 2018, and adopted by the International Union for the Conservation of Nature).

Salt Flat – Unvegetated seasonal wetlands that fluctuate between dry, hypersaline conditions and shallow freshwater and/or tidal inundation (Briere 2000, Yechieli and Wood 2002).

Sea-level Rise Migration Zone – Inland land required to accommodate marsh migration with sea-level rise.

Soft Substrate – Unconsolidated sediments in shallow subtidal areas of estuaries with less than 10% colonization by Submerged Aquatic Vegetation (SAV).

Southern California Bight – The Bight is a distinct bioregion of California which extends from Point Conception in Santa Barbara County to Punta Banda, South of Ensenada, in Baja California, Mexico and includes the marine-coastal interface and the coastal wetlands and watersheds. For the RSU, the Bight's boundaries are from Point Conception to the U.S.-Mexico border at Tijuana.

Southern California Wetland Recovery Project (WRP) – The WRP is a broad-based collaboration, led by the California Natural Resources Agency and supported by the California State Coastal Conservancy and Earth Island Institute, that has public agencies, non-profits, scientists, and local communities working cooperatively to acquire and restore rivers, streams, and wetlands in coastal Southern California.

Stream Order – A measure of the relative size of streams with the smallest tributaries referred to as first-order streams, up to a twelfth-order waterway. First- through third-order streams are called headwater streams.

Submerged Aquatic Vegetation – Rooted, vascular plants that grow completely underwater except for periods of brief exposure at low tides.

Subregion – Regions along the Bight that are differentiated by significant physiographic characteristics such as steepness of the watershed and width of coastal plain. Geopolitical boundaries may also influence divisions between subregions. The Science Advisory Panel divided the Bight as such:
- Santa Barbara Subregion (Point Conception to the Santa Barbara/Ventura County Line)
- Ventura Subregion (Santa Barbara/Ventura County Line to Santa Monica Mountains)
- Santa Monica Subregion (Santa Monica Mountains to Palos Verdes)
- San Pedro Subregion (Palos Verdes to San Onofre Mountains)
- San Diego Subregion (San Onofre Mountains to Tijuana)

Tidal Prism – Measure of the amount of water entering the estuary on each tide when it is connected to the ocean.

Tidal Extent – Farthest point upstream where a river is affected by tidal fluctuations.

Tidal Range – Vertical difference between the high tide and the succeeding low tide.

Unvegetated Flat – Areas of coastal wetland that are unvegetated (i.e., salt flats & mudflats).

Vegetated Marsh – Areas of coastal wetland dominated with wetland plants.

Vision – The "big picture" philosophy or guiding principle of the WRP. A vision is the ideal view of the future that we are trying to achieve, no matter how unlikely they may seem now, such that over time the sum of projects add up to a greater whole. Broader and more general than an objective or an outcome.

Watershed – Land area that channels rainfall and snowmelt to creeks, streams, and rivers, and eventually to outflow points such as reservoirs, bays, and the ocean (https://oceanservice.noaa.gov/facts/watershed.html).

Wetland Advisory Group (WAG) – The Wetlands Advisory Group (WAG) is a committee of Southern California wetland resource managers. This group is an iteration of the County Task Forces (see above), with representatives from all 5 counties in the WRP region, who were selected based on their unique knowledge of local wetlands.

Wetland Managers Group (WMG) – The Wetland Managers Group (WMG) consists of staff members from the 18 state and federal agencies that make up the WRP. The group meets on a monthly basis to make vital decisions that guide our programmatic and project goals. They facilitate interagency coordination and work collaboratively to identify a set of projects for the Work Plan and other activities to implement the WRP Regional Strategy.

Wetland System – Encompasses coastal wetlands (see above) plus additional habitat such as deeper subtidal areas. Systems can also include fragmented tidal wetlands that were historically part of a connected wetland but are currently functioning as individual wetlands.

Wetland-Upland Ecotone – A narrow band of habitat where wetlands and uplands meet that contains vegetations types from both habitats. The ecotone boundaries are set by factors such as soil salinity and moisture (Callaway et al. 1990; James and Zedler 2000).

Wetland-Upland Transition Zone – The wetland-upland transition zone (t-zone) connects tidal wetlands to adjacent terrestrial habitats, providing flooding refuge for wildlife, space to accommodate marsh transgression with sea-level rise, and other vital ecosystem functions. The ecosystem services associated with the transition zone occur within a range of tens to hundreds of yards. The width of the t-zone varies substantially by location, and can be defined more or less narrowly depending on the ecosystem services being looked at. Mapping an upper boundary of potential t-zone helps identify opportunities for protection and restoration. The location of the upper boundary will be influenced by topography, differing between areas with gradual slopes ("hillslope"), areas with cliff/bluffs, and for streams and rivers and can be up to 1,600 ft wide. The lower t-zone boundary is approximated using mean higher high water (MHHW).

Work Plan – Current list of projects that the WRP partner agencies have identified as high priorities for funding and technical assistance. Projects are added to the Work Plan through an application and vetting process involving review by the Wetland Managers Group and adoption by the Directors Group.

References

Ambrose, R.F., J.C. Callaway, and S.F. Lee. 2006. An evaluation of compensatory mitigation projects permitted under Clean Water Act Section 401 by the California State Water Quality Control Board, 1991-2002. California Environmental Protection Agency, California State Water Resources Control Board Los Angeles Region (Region 4). Los Angeles (CA).[Internet]. Contract, (03-259), pp.250-0.

Ballard, J., J. Pezda, and D. Spencer. 2016. An economic valuation of Southern California coastal wetlands. University of California, Santa Barbara, Master's Thesis Group Project. http://www.esm.ucsb.edu/research/2016Group_Projects/documents/SoCalWetlands_FinalReport.pdf

Beller, E.E., R.M. Grossinger, M.N. Salomon, S.J. Dark, E.D. Stein, B.K. Orr, P.W. Downs, T.R. Longcore, G.C. Coffman, A.A. Whipple, R.A. Askevold, B. Stanford, and J.R. Beagle. 2011. Historical ecology of the lower Santa Clara River, Ventura River, and Oxnard Plain: an analysis of terrestrial, riverine, and coastal habitats. Prepared for the State Coastal Conservancy. A report of SFEI's Historical Ecology Program, SFEI Publication #641, San Francisco Estuary Institute, Oakland, CA.

Beller, E.E., S.A. Baumgarten, R.M. Grossinger, T.R. Longcore, E.D. Stein, S.J. Dark, and S.R. Dusterhoff. 2014. Northern San Diego County Lagoons historical ecology investigation: regional patterns, local diversity, and landscape trajectories. SFEI Publication #722. Prepared for the State Coastal Conservancy.

Bernstein, B., K. Merkel, B. Chesney, and M. Sutula. 2011. Recommendations for a southern California regional eelgrass monitoring program. Southern California Coastal Water Research Project Technical Report 632.

Biology Online. 2018. "Ecosystem function." Biology Online Dictionary. https://www.biology-online.org/.

Briere, P. R. 2000. Playa, playa lake, sabkha: Proposed definitions for old terms. Journal of Arid Environments 45:1-7.

PAD (California Fish Passage Assessment Database). 2018. California Department of Fish and Wildlife: Bios Interactive Web Map. https://map.dfg.ca.gov/bios/?al=ds69.

CWQMC (California Water Quality Monitoring Council). 2016. Elements of wetland and riparian area monitoring plan (WRAMP). https://mywaterquality.ca.gov/monitoring_council/wetland_workgroup/wramp/index.html. Accessed August 2018.

California Wetland Interagency Team. 2017. California Wetland Program Plan, 2017-2022. https://www.waterboards.ca.gov/water_issues/programs/cwa401/docs/ca_wetland_program_plan_2017_2022_signed.pdf

CWMW (California Wetland Monitoring Workgroup). 2008. California's Wetland Demonstration Program Pilot. A Final Project Report to the California Resources Agency. MA Sutula, JN Collins, R Clark, RC Roberts, ED Stein, CS Grosso, A Wiskind, C Solek, M May, K O'Connor, AE Fetscher, JL Grenier, S Pearce, A Robinson, C Clark, K Rey, S Morrissette, A Eicher, R Pasquinelli, K Ritter. Technical Report 572. Southern California Coastal Water Research Project. Costa Mesa, CA.

CWMW (California Wetland Monitoring Workgroup). 2010. Tenets of a state wetland and riparian monitoring program (WRAMP). 67 pages. http://www.mywaterquality.ca.gov/monitoring_council/wetland_workgroup/docs/2010/tenetsprogram.pdf

CWMW (California Wetland Monitoring Workgroup). 2014. California Aquatic Resources Status and Trends Program: Mapping Methodology. E.D. Stein, J.S. Brown, K. Cayce, M. Klatt, M. Salomon, P. Pendleton, S. Dark, K. O'Connor, and C. Endris. Technical Report 833. Southern California Coastal Water Research Project Authority. Costa Mesa, CA.

CWMW (California Wetland Monitoring Workgroup). 2018. EcoAtlas. Accessed 2018. https://www.ecoatlas.org.

Callaway, R. M., S. Jones, W. R. Ferren Jr, and A. Parikh. 1990. Ecology of a mediterranean-climate estuarine wetland at Carpinteria, California: plant distributions and soil salinity in the upper marsh. Can. J. Bot 68:1139–1146.

City of Santa Ana. 2006. Santa Ana River Task Force vision plan: history and historical timeline. Accessed on May 23, 2018: http://www.ci.santa-ana.ca.us/parks/RiverVisionPlan.asp.

Collins, J. and E. Stein. 2018. Wetland and stream rapid assessments. Amsterdam(NL): Elsevier. Chapter 4, California Rapid Assessment Method for Wetlands and Riparian Areas (CRAM).

Costanza, R., R. d'Arge, R. De Groot, S. Farber, M. Grasso, B. Hannon, K. Limburg, S. Naeem, R.V. O'neill, J. Paruelo, and R.G. Raskin. 1997. The value of the world's ecosystem services and natural capital. Nature, 387(6630): 253.

Cowardin, L.M., V. Carter, F.C. Golet, and E.T. LaRoe. 1979. Classification of wetlands and deepwater habitats of the United States. U.S. Fish and Wildlife Service Report No. FWS/OBS/-79/31. Washington, D.C.

Daily, G.C., S. Alexander, P.R. Ehrlich, L. Goulder, J. Lubchenco, P.A. Matson, H.A. Mooney, S. Postel, S.H. Schneider, D. Tilman, and G.M. Woodwell. 1997. Ecosystem services: Benefits supplied to human societies by natural ecosystems. ESA Issues in Ecology (2):2-16.

Dark, S., E.D. Stein, D. Bram, J. Osuna, J. Monteferante, T. Longcore, R. Grossinger, and E. Beller. 2011. Historical ecology of the Ballona Creek Watershed. Southern California Coastal Water Research Project Technical Report # 671.

Davies, B.R. 1982. Studies on the zoobenthos of some southern Cape coastal lakes. Spatial and temporal changes in the benthos of Swartvlei, South Africa, in relation to changes in the submerged littoral macrophyte community. Journal of the Limnological Society of Southern Africa 8: 33–45.

Egan, D., and E.A. Howell, editors. 2001. The Historical Ecology Handbook: a Restorationist's Guide to Reference Ecosystems. Island Press, Washington. D. C., USA

Fejtek, S., M. Gold, G. MacDonald, D. Jacobs, and R. Ambrose. 2014. Best management practices for southern California coastal wetland restoration and management in the face of climate change. University of California Los Angeles, Institute of the Environment and Sustainability.

Fejtek, S.M. 2017. The implications of current restoration practices and regulatory policy for recovery of the federally endangered southern california steelhead (Order No. 10636892). Available from ProQuest Dissertations & Theses Global. (1965459912). Retrieved from https://search.proquest.com/docview/1965459912?accountid=14512

Gittman R.K., S.B. Scyphers, C.S. Smith, I.P. Neylan, and J.H. Grabowski. 2015. Ecological consequences of shoreline hardenings: A meta-analysis. Bioscience 66(9): 763–773.

Gleason, M.G., S. Newkirk, M.S. Merrifield, J. Howard, R. Cox, M. Webb, J. Koepcke, et al. 2011. A conservation assessment of west coast (USA) estuaries. The Nature Conservancy.

Griggs, G., J. Árvai, D. Cayan, R. DeConto, J. Fox, H.A. Fricker, R.E. Kopp, C. Tebaldi, E.A. Whiteman(California Ocean Protection Council Science Advisory Team Working Group). 2017. Rising Seas in California: An Update on Sea-Level Rise Science. California Ocean Science Trust, April 2017.

Hawley, R.J., B.P. Bledsoe, E.D. Stein, and B.E. Haines. 2012. Channel evolution model of semiarid stream response to urban-induced hydromodification. Journal of the American Water Resources Association (JAWRA) 48(4): 722-744. DOI: 10.1111/j.1752-1688.2012.00645.

Hruby, T. 2013. Update on wetland buffers: the state of the science, final report, October 2013. Washington State Department of Ecology Publication #13-06-11.

IEP (Interagency Ecological Program). 2012. IEP Science Strategy – Needs for near-term science In Five Areas of Emphasis. 184 pages. https://www.water.ca.gov/-/media/DWR-Website/Web-Pages/Programs/Environmental-Services/Interagency-Ecological-Program/Files/A-Historical-Perspective-of-the-Interagency-Ecological-Program.pdf?la=en&hash=9F2AA99B4F5709C270A70B0137807DEE8A7EF740

Jacobs, D., E.D. Stein, and T. Longcore. 2011. Classification of California estuaries based on natural closure patterns: Templates for restoration and management. Technical Report 619.a. SCCWRP.

James M.L. and J.B. Zedler. 2000. Dynamics of wetland and upland subshrubs at the salt marsh-coastal sage scrub ecotone. BioOne 143(2): 298-311.

Johnston, K., I. Medel, S. Anderson, E. Stein, C. Whitcraft, and J. Crooks. 2015. California estuarine wetland monitoring manual (level 3). Prepared by The Bay Foundation for the United StatesEnvironmental Protection Agency. 297 pages.

Keddy, P.A., L.H. Fraser, A.I. Solomeshch, W.J. Junk, D.R. Campbell, M.T. Arroyo, and C.J. Alho. 2009. Wet and wonderful: the world's largest wetlands are conservation priorities. BioScience. 59(1): 39-51.

Kennish, M.J., (Ed.), 2004. Estuarine research, monitoring, and resource protection. CRC Press, Boca Raton, Fla.

Millenium Ecosystem Assessment. 2005. Ecosystems and human well-being: current state and trends. Millennium Ecosystem Assessment, Global Assessment Reports. https://www.millenniumassessment.org/documents/document.356.aspx.pdf.

Moreno-Mateos, D., M.E. Power, F.A. Comín, and R. Yockteng. 2012. Structural and functional loss in restored wetland ecosystems. PLoS biology. 10(1): e1001247.

National Research Council. 2012. Sea-level rise for the coasts of California, Oregon, and Washington: Past, present, and future. Page (R. A. Dalrymple and et al., Eds.) Lighthouse. National Academy of the Sciences, Washington, DC.

Needles L.A., S.E. Lester, R. Ambrose, A. Andren, M. Beyeler, M.S. Connor, J.E. Eckman, B.A. Costa-Pierce, S.D. Gaines, K.D. Lafferty, H.S. Lenihan, J.P. Peterson, M.S. Peterson, A.E. Scaroni, J.S. Weis, and D.E. Wendt. 2015. Managing Bay and Estuarine Ecosystems for Multiple Services. Estuaries and Coasts 38: 35–48.

Perissinotto, R., D.D. Stretch, A.K. Whitfield, J.B. Adams, A.T. Forbes, and N.T. Demetriades. 2010. Temporarily open/closed estuaries in South Africa. New York: Nova Science Publishers.

Regional Strategy (Southern California Wetlands Recovery Project). 2001. Southern California Wetlands Recovery Project Regional Strategy. http://scwrp.org/wp-content/uploads/2013/08/WRP-Regional-Strategy.pdf

Riddin, T., and J.B. Adams. 2008. Influence of mouth status and water level on the macrophytes in a small temporarily open/closed estuary. Estuarine, Coastal and Shelf Science 79: 86–92.

Schumm, S.A., M.D. Harvey, and C.C. Watson. 1984. Incised channels: morphology, dynamics, and control. Water Resources Publications, Littleton, Colorado, 200 pp

SEQ (The SEQ Ecosystem Services Framework). n.d. http://www.ecosystemservicesseq.com.au/index.html.

SER (Society for Ecological Restoration). 2018. Ecological restoration. https://www.ser.org/default.aspx

Stein, E.D., S. Dark, T. Longcore, N. Hall, M. Beland, R. Grossinger, J. Casanova, and M. Sutula. 2007. Historical ecology and landscape change of the San Gabriel River and floodplain. Southern California Coastal Water Research Project Technical Report #499.

Stein, E.D., K. Cayce, M. Salomon, D.L. Bram, D. De Mello, R. Grossinger, and S. Dark. 2014. Wetlands of the Southern California Coast - historical extent and change over time. SCCWRP Technical Report 826, SFEI Report 720, Southern California Coastal Water Research Project, San Francisco Estuary Institute, California State University Northridge Center for Geographical Studies.

Stillwater Sciences. 2011. Geomorphic assessment of the Santa Clara River watershed: synthesis of the lower and upper watershed studies, Ventura and Los Angeles counties, California. Prepared by Stillwater Sciences, Berkeley, California for Ventura County Watershed Protection District, Los Angeles County Department of Public Works, and the U.S. Army Corps of Engineers–L.A. District.

Strahler, A.N. 1952. Hypsometric (area-altitude) analysis of erosional topology. Geological Society of America Bulletin 63 (11): 1117–1142.

Stralberg, D., M. Brennan, J.C. Callaway, J.K. Wood, L.M. Schile, D. Jongsomjit, M. Kelly, V.T. Parking, and S. Crooks. 2011. Evaluating tidal marsh sustainability in the face of sea-level rise: A hybrid modeling approach applied to San Francisco Bay. PLoS ONE 6(11): e27388. https://doi.org/10.1371/journal.pone.0027388

Sutula, M., J.N. Collins, R. Clark, C. Roberts, E. Stein, C. Grosso, A. Wiskind, C. Solek, M. May, K. O'Connor, A.E. Fetscher, J.L. Grenier, S. Pearce, A. Robinson, C. Clark, K. Rey, S. Morrissette, A. Eicher, R. Pasquinelli, and K. Ritter. 2008. California's Wetland Demonstration Program Pilot: a final draft project report for review to the California Resources Agency. Technical Report 572. Southern California Coastal Water Research Project, Costa Mesa. ftp://ftp.sccwrp.org/pub/download/DOCUMENTS/TechnicalReports/572_WDP.pdf

Sweeney, B.W., and J.D. Newbold. 2014. Streamside forest buffer width needed to protect stream water quality, habitat, and organisms: a literature review. JAWRA Journal of the American Water Resources Association. 50(3): 560-584. DOI: 10.1111/jawr.12203.

U.S. EPA, 2013. California integrated assessment of watershed health. EPA publication 841-R-14-003.

Walker, B., C.S. Holling, S.R. Carpenter, and A. Kinzig. 2004. Resilience, adaptability and transformability in social–ecological systems. Ecology and Society 9(2): 5. http://www.ecologyandsociety.org/vol9/iss2/art5/

Wenger, S., 1999. A review of the scientific literature on riparian buffer width, extent and vegetation.

Werner, K.J., and J.B. Zedler. 2002. How sedge meadow soils, micro-topography, and vegetation respond to sedimentation. Wetlands 22: 451-466

Whitfield, A.K. 1984. The effects of prolonged aquatic macrophyte senescence on the biology of the dominant fish species in a southern African coastal lake. Estuarine, Coastal and Shelf Science 18: 315–329.

Yechieli, Y., and W.W. Wood. 2002. Hydrogeologic processes in saline systems: playas, sabkhas, and saline lakes. Earth-Science Reviews 58:343-365.

Zedler, J.B., and S. Kercher. 2004. Causes and consequences of invasive plants in wetlands: Opportunities, opportunities, and outcomes. Critical Reviews in Plant Sciences 23(5): 431-452.

www.ingramcontent.com/pod-product-compliance
Lightning Source LLC
Chambersburg PA
CBHW041729210326
41598CB00008B/821